World Water Resources

Volume 7

Series Editor
V.P. Singh, Department of Biological and Agricultural Engineering & Zachry
Department of Civil Engineering, Texas A&M University, College
Station, TX, USA

This series aims to publish books, monographs and contributed volumes on water resources of the world, with particular focus per volume on water resources of a particular country or region. With the freshwater supplies becoming an increasingly important and scarce commodity, it is important to have under one cover up to date literature published on water resources and their management, e.g. lessons learnt or details from one river basin may be quite useful for other basins. Also, it is important that national and international river basins are managed, keeping each country's interest and environment in mind. The need for dialog is being heightened by climate change and global warming. It is hoped that the Series will make a contribution to this dialog. The volumes in the series ideally would follow a "Three Part" approach as outlined below: In the chapters in the first Part *Sources of Freshwater* would be covered, like water resources of river basins; water resources of lake basins, including surface water and under river flow; groundwater; desalination and snow cover/ice caps. In the second Part the chapters would include topics like: *Water Use and Consumption*, e.g. irrigation, industrial, domestic, recreational etc. In the third Part in different chapters more miscellaneous items can be covered like impacts of anthropogenic effects on water resources; impact of global warning and climate change on water resources; river basin management; river compacts and treaties; lake basin management; national development and water resources management; peace and water resources; economics of water resources development; water resources and civilization; politics and water resources; water-energy-food nexus; water security and sustainability; large water resources projects; ancient water works; and challenges for the future. Authored and edited volumes are welcomed to the series. Editor or co-editors would solicit colleagues to write chapters that make up the edited book. For an edited book, it is anticipated that there would be about 12-15 chapters in a book of about 300 pages. Books in the Series could also be authored by one person or several co-authors without inviting others to prepare separate chapters. The volumes in the Series would tend to follow the "Three Part" approach as outlined above. Topics that are of current interest can be added as well.

Readership
Readers would be university researchers, governmental agencies, NGOs, research institutes, and industry. It is also envisaged that conservation groups and those interested in water resources management would find some of the books of great interest. Comments or suggestions for future volumes are welcomed.

Series Editor:
V.P. Singh, Department of Biological and Agricultural Engineering & Zachry Department of Civil Engineering, Texas A&M University, TX, USA.
Email: vsingh@tamu.edu

More information about this series at http://www.springer.com/series/15410

Amin Shaban

Water Resources of Lebanon

 Springer

Amin Shaban
National Council for Scientific Research
Beirut, Lebanon

ISSN 2509-7385   ISSN 2509-7393   (electronic)
World Water Resources
ISBN 978-3-030-48716-4   ISBN 978-3-030-48717-1   (eBook)
https://doi.org/10.1007/978-3-030-48717-1

This Springer imprint is published by the registered company Springer Nature Switzerland AG
The registered company address is: Gewerbestrasse 11, 6330 Cham, Switzerland

Foreword

Dr. Amin Shaban, building on more than 30 years of research in the field, and working at the National Council for Scientific Research – Lebanon (CNRS-L) provides an insightful look and the first comprehensive book on water resources and its management in Lebanon.

It has been a pleasure to have witnessed, first hand, the contribution Amin Shaban has made and the evident mark he continues to create in his field of water research both nationally and regionally. Building on years of research, frequent field work and promising collaborative partnerships, Amin Shaban has contributed widely to various water-related issues with scientific knowledge, innovative analysis techniques and quality publications.

This well-illustrated 10-chapter book provides a timely endeavour to reflect on available studies, complement the research with innovative and new techniques for analysis, and provide updated and new estimations of water resources in Lebanon, which encompass various sources, including rivers, springs, snow, lakes, wetlands and groundwater rock formations, within a holistic approach that includes territorial and atmospheric variables, demand and supply consideration, as well as affecting topographic and geological features.

Located within the water-scarce region of the Middle East, Lebanon stands out with its diverse topography, encompassing various and rich water resources. Yet despite this, the challenges lie in increasing water shortages, inadequate water estimations and an increasing demand on a limited water supply.

Building on the rich data provided in his book, the author's later chapters help highlight existing challenges that exasperate the stress on surface water and groundwater resources in Lebanon. With both natural and anthropogenic challenges, that is, topography, climate, hydrology and shared water resources, exasperated by human interference, that is, population growth, water quality deterioration and unwise use of a limited resource, the author declares that "Water in Lebanon is in jeopardy".

But the book doesn't end there. It continues to provide, based on a scientific approach, outlines for active solutions built on expertise, observations and a scientific outline for further actions within a study that encompasses aspects of surface water harvesting, technologies of artificial groundwater recharge, tapping ground water discharges into the sea, reducing water contamination, proposed economic policies to enhance the water sector, ethics and moral behaviour, as well as mainstreaming the applicability of Sustainable Development Goals in water (particularly SDG-6).

It is my conviction that the data and analyses provided by Amin Shaban will be extremely beneficial to decision makers, stakeholders and beneficiaries, providing a strong scientific footing for mitigation policies and sustainable development of the water sector in Lebanon, whilst also providing a comprehensive case study with various dimensions for further regional and international applicability.

Secretary General   Mouïn Hamzé
National Council for Scientific Research
Beirut
Lebanon

Preface

Water shortage remains the most crucial geoenvironmental issue in many regions of the world, especially where annual precipitation is very low. Recently, this issue has been addressed aggressively due to existing challenges. Moreover, water resources are under serious threat because of increase in water demand. Therefore, water has become a valuable commodity and also results in geopolitical conflicts between neighbouring nations. In Lebanon, there is a severe water crisis and demand for water has been exacerbated. However, no improvement has been observed in the water sector since 1990.

The climate of Lebanon is relatively wet, even though some spells of dryness occur. This is well evidenced by the green cover that spreads over more than 80% of the Lebanese territory.

In Lebanon, several observations point to abundant water resources, spanning from watercourses on terrain surface to a large number of springs and wetlands. In addition, snow remains on the Lebanese mountains for a few months and may extend from one year to another. Moreover, groundwater potentiality is feasible in large part of the Lebanese territory. It is, therefore, a paradox that the country with abundant water supply is facing water shortage, and the public water supply fulfils only 35% of water demand. Besides, there are many studies, projects and initiatives done to assure water supply, but still there is a water shortage. It is also surprising enough that the funding by the Lebanese government does not exceed 10% of financial resources to conduct studies and assess water resources, and this might be the case for the halted projects. This, in turn, raises several questions about the management approaches and about the concern of the Lebanese government in securing water, a vital element, for its inhabitants.

Several initiative scan be taken to rescue water resources and to provide pure and sufficient quantity of water for inhabitants. For example, but not limited to, if shared water resources of Lebanon are well managed, 80% of water demand can be met. Also, if rainwater harvesting is implemented, even on individual basis by building small check dams, approximately more than 3 million m^3 of surface water can be conserved. Moreover, if simplified methods like rooftop harvesting is adopted, there will be an increase of about 18 m^3/capita/year. Therefore, solutions exist, but proper execution is still lacking.

Several research are being conducted out to assess water resources in Lebanon, and most of these studies aim to detect the reasons behind water stress in the country. It is frustrating to find that many studies attribute the failure in the water sector in Lebanon to physical challenges, with special emphasis on climate change, but overlook the poor water management.

In conclusion, "Water in Lebanon is in jeopardy". Hence, practical actions should be taken, and the government must dedicate much concern to the water sector by employing experts in water resources and from related disciplines to carry out creditable studies.

If we envision the future of water in Lebanon, all scenarios seem pessimistic. This emergency has been declared even by decision-makers and people who manage the water sector. Therefore, there is no more time to waste and rapid action is required.

As the author of this book, I could build a vision on current and future trends in water sector. My experience and knowledge on water resources in Lebanon, including in-depth investigation, research and field measures accompanied with the use of advanced techniques, were useful to produce this comprehensive book.

This book describes all aspects of water resources in Lebanon, the surface and subsurface ones, with detailed discussion on new estimations. This book discusses all physical and anthropogenic factors that influence water resources. It also underpins comprehensive discussion on rivers, springs, snow, lakes, reservoirs, wetlands and groundwater. It, eventually, presents the existing challenges and proposes possible solutions to overcome them. Therefore, this book would be a helpful tool for different-level stakeholders, starting from individuals to high-level decision-makers.

The National Council for Scientific Research of Lebanon (CNRS-L) has shown concern to water resources studies. Therefore, CNRS-L always cooperates with and helps researchers to perform various water studies, especially those using advanced techniques for analysis. Considering the important objectives and scientific relevance of water studies, CNRS-L introduces facilities and logistics to me to author this book.

Beirut, Lebanon   Amin Shaban

Contents

Acronyms

CDR	Council of Development and Reconstruction
CESBIO	Centre d'Etudes Spatiales de la Biosphère
CNRS–L	National Council for Scientific Research–Lebanon
COMEST	World Commission on the Ethics of Science and Technology
CPA	Consumer Protection Association
CS	Council for South
FAO	Food and Agriculture Organization
GDEM	Global Digital Elevation Model
GIS	Geographic Information System
IETC	International Environmental Technology Centre
IHP	International Hydrology Programme
IRD	Institut de Recherche pour le Développement – France
IWMI	International Water Management Institute
LARI	Lebanese Agronomical Research Institute
LRA	Litani River Authority
MoA	Ministry of Agriculture
MoE	Ministry of Environment
MoEW	Ministry of Energy and Water
MoFA	Ministry of Foreign Affairs
NWSS	National Water Sector Strategy
OECD	Organization for Economic Cooperation and Development
SDGs	Sustainable Development Goals
SNC	Second National Communication for Lebanon
UNDP	UN Development Programme
UNEP	United Nations Environmental Program
UN-HABITAT	The United Nations Human Settlements Programme
UNHCR	UN High Commissioner for Refugees
UNIECF	United Nations International Children's Emergency Fund
USAID	United States Agency for International Development
WB	World Bank
WE	Water Establishments
WEF	World Economic Forum

Chapter 1
Introduction

Abstract Water resources in the Middle East Region, where they are scarce, is a matter of utmost significance. However, it must be made clear that Lebanon has a diverse topography that makes it with different physical setting from the surrounding regions in the Middle East. Even though, Lebanon has a small area (10,452 km^2), yet it encompasses different aspects of water resources whether on surface including rivers, springs, snow and lakes; and sub-surface where a number of aquiferous rock formations and karstic conduits exist with considerable amounts of water. Nevertheless, there is still complain about water supply/demand. The country becomes under water stress and suffering from water shortage. Meanwhile, creditable estimations on water resources are still inadequate. Recently, challenges on water resources have been exacerbated including the population growth and the increased water demand, plus the changing climatic conditions. The existing management approaches done by the governmental sector are few enough to adapt water sector to these challenges. Thus, managing the demand of water in Lebanon is substantially adopted by the individuals rather than the public sector itself. This chapter will introduce an overview on the Lebanon's territory, then it will illustrate different measurements on water resources including mainly water availability, demand and supply. In addition, an inventory on the previous obtained studies will be mentioned.

Keywords Water stress · Mountainous region · River flow · Contamination · Eastern Mediterranean

1.1 Lebanon in the Regional View

Lebanon, the Middle East country along the Eastern Mediterranean Sea, is was ranked as the 162nd country worldwide, and the 19th Arab country in terms of the geographic area. Lebanon is mainly a mountainous region where a number of elevated areas occur and represent a chain of mountains extending parallel to the

Mediterranean. The country occupies the highest crest in the entire Middle East Region, where a 3088 m altitude is.

Lebanon has a relatively small area (10,452 km²) with a maximum coastal length of 220 km and 85 km width. Even though Lebanon has limited geographic area, yet it encompasses diverse physiography which is remarkably different from the surrounding regions. Thus, three major geomorphological features exist. They are the Mount-Lebanon, Bekaa Plain and the Anti-Lebanon. These features, with their exposed rock bodies, form a climatic barrier that captures wet air masses blown by wind from the Mediterranean to the east, and this is the main reason why Lebanon is characterized by relatively high precipitation rate including rainfall and snow.

The entire territory of Lebanon is considered as a regional water junction where surface water flows from Lebanese territory into three regional drainage systems. These systems are: (1) the occidental Lebanon drainage system which comprises a number of relatively short rivers and streams, and thus provides water into the Mediterranean Sea, (2) the northern Bekaa Plain which is represented by the Al-Assi River as a major tributary of the Orontes River that spans to Syria and then Turkey, and (3) the Hermoun Mountain which is represented by the Hasbani-Wazzani River a major tributary of the Jordan River (Shaban and Hamzé 2017).

The geographic location of Lebanon implies the following geographic coordinates:

33° 03′ 14″ N and 34° 41′ 32″N &
35° 06′ 14″ E and 36° 37′ 25″E

Lebanon is almost situated in a semi-arid region, but its morphology characterized it by relatively wet climate. Even though, higher temperatures are noticed as a regional climatic phenomenon, yet a recent study depended on advanced statistical analysis shows that Lebanon is a located in a sub-humid-climatic zone (Shaban and Houhou 2015).

The geographic location of Lebanon, which is almost adjacent to the borders of three continents (Africa, Asia and Europe), often put it within diverse regional geographical nomenclatures. Thus, Lebanon belongs to the regions of: Middle East (ME), Middle East-North Africa (MENA), Arab Region (AR), and ESCWA Region.

Due to its remarkable nature and setting, Lebanon is always viewed as a distinguished place, and thus it was named by several descriptions, such as: the Swiss of the Levant, Water Tower of the Middle East, Cedars Land, etc.

1.2   Historical Overview

Even though it has a small geographic area, yet Lebanon is one of rare countries which encompass the entire elements of the water cycle (e.g. snow cover, water-bearing conduits, renewable groundwater, sub-marine springs, etc.). In this view, and since the ancient time periods, Lebanon is known by abundant water resources, and the remarkable thing is that these resources have diverse aspects whether on surface or the sub-surface. This made Lebanon a country with distinguished historical water-related constructions, such as water collecting ponds, man-made cannels, terraces, watermills, caves and stone bridges (Fig. 1.1).

In this regards, many appreciations and poems celebrated Lebanon's landscape throughout ancient history, and water was the main element of the charming nature. This includes snow cover, running water in rivers and springs, water-bearing grottos. These aspects of water resources result colourful fertile lands.

Fig. 2.1 Old bridge on Naher El-Kaleb. Photo taken in 1810 by Sami Toubia (Maïla-Ateihe 2009)

The history of water in Lebanon has been come to light since the Ottoman Empire (1516–1919). During this period, the empire undertook a number of reforms on the basic services including mainly the drinking water and land. Therefore, several regulations and diligences have been put on water and land management, such as: *amirié land, tariffs, firman*, etc. (Mallat 2003).

In this respect, the most significant established regulation was the bring water from Naher El-Kalb to Beirut and provide running water to the city (OEB 1996). This affected many agricultural lands, notably where irrigation canals and watermills exist. Thus the issue arose of recognizing the water rights acquired by dignitaries and monasteries. Thus, the concession changed the distribution of water access by bringing together various social groups outside of the established framework based on the confessional system.

This episode of Ottoman reforms had a major influence on the water sector. The identification of legal problems linked to this concession and the solutions adopted

Fig. 1.1 Location of Lebanon

strongly influenced the Ottoman Civil Code which is assigned a "Mecelle" (Ghiotti and Riachi 2013).

The civil war (1975–1990) occurred in Lebanon was one of the most hindered factors in the development of the water sector. Therefore, new water projects and supply implementations have been proposed to form a new era in the water policy of Lebanon.

Yet, Lebanon has only two large dams (i.e. Qaraaoun and Shabrouh) while the country encompasses more than 40 major watercourses that run water for a couple of months per year. This reflects the shortage in applying proper and required water measures for better water resources management.

Nowadays, it can be tedious to match the reality with the history of natural resources in Lebanon, notably when it comes to water. Therefore, access to water became a geo-environmental issue, and the majority of consumers only have access to water for few days per month. Hence, water provided by the public water sector is very limited, and therefore, consumers depend on themselves to access water for different purposes. This in turn led to uncontrolled behaviours to reach water resources, such as digging chaotic boreholes, direct pumping from rivers and springs, etc.

Recently, water has become a commodity for trading in Lebanon. There is the bottled water for more than 40 trademarks, water tankers and tractor are commonly seen, in addition boreholes are widespread, notably the illegal ones. For example, there are over 30.000 illegal groundwater wells dug in the Greater Beirut and the adjacent mountainous region. In the same time, water contamination became a daily issue for discussion, but no solutions have been reached yet. In conclusion, water resources in Lebanon are in jeopardy.

1.3   General Water Measurements

Usually, Lebanon is described as the country with plenty water resources, and perhaps this is because water can be seen everywhere on the Lebanese territory. However, the supply/demand reflects a contradictory figure, and the overall figure on water reserves and the mechanism of water retention is still obscure.

Therefore, the condensed cold air masses from the Mediterranean result considerable precipitation if compared with the precipitation in the neighboring Middle East countries. Hence, general estimates reveal that the average annual rainfall rate in Lebanon ranges between 700 mm and 1500 mm, and snow covers annually more 2000–2500 km². Moreover, Lebanon occupies 14 perennial watercourses (i.e. rivers) and more than 1500–2000 springs with permanent flow. In addition, there are a number of aquiferous formations and karstic conduits which store considerable volume of groundwater.

There are many estimates on the renewable water resources in Lebanon, which have been plotted by many sources, such as: UNDP and FAO 1983, Jaber 1995, Bou Zeid and El-Fadel 2002, Fawaz 2007, and the National Council for Scientific Research (CNRS-L) 2015. However, renewable water resources in Lebanon have been subjected to abrupt fluctuations from one year to another as a result of climatic variability. As an example, a recent wet storm hit Lebanon in the beginning of 2019 for a couple of days; therefore, this storm added about 1.5 billion m^3 of water which is equivalent to the amount of stored groundwater in Lebanon.

The average volume of the precipitated water in Lebanon is about 9.5 billion m^3/year, and after subtracting the evapotranspiration, which is about 4.85 billion m^3/year, or equivalent to 51% of the precipitated water, the rest will be 4.65 billion m^3/year. However, some studies stated that the renewable water resources in Lebanon is averaging about 4.1 billion m^3/year (CESFB 2018).

1.4 Water Availability

In general, water resources are considered as plenty since they can be observed everywhere in Lebanon, notably the surface resources. In addition, the tangible wet climate over couple of months (including the snow cover and rainfall) brings good sense towards water availability. This optimistic figure has been assured since water availability exceeds the threshold of the standard water-poverty which is determined at 1000 m^3/capita/year, even though the supply and demand are still imbalanced. Hence, several estimates done to calculate the water quota for inhabitants in Lebanon.

Lonergan and Brooks (1994), for example, calculated water availability per capita in Lebanon at 950 m^3/year. While, Shahin (1996) estimated much higher value of approximately 3750 m^3/year. However, recent estimations were done by Shaban (2011) where detailed socioeconomic survey has been applied to different regions in Lebanon including the estimation of water availability and even the consumption rates. In the obtained survey, Shaban considered all available resources, virtual water, climatic variability and oscillations as well as diversity of water use in different regions. Thus, the resulted estimation was at 1350 m^3/capita/year.

Nevertheless, the rate of water availability has been declined and it was believed that the changing climate is the main reason behind, but this understanding was no longer convince when the demographic control was involved after the year 2015 due to the displacement of large number of people (estimated at 2 million people) from the surrounding countries as a result of political conflicts in the Middle East Region. Therefore, it was lately estimated at 921 m^3/capita/year (Shaban 2016). This means that water availability in Lebanon has been decreased at about 429 m^3/capita/year (i.e. equivalent to 32%) between 2011 and 2016.

1.5 Water Demand

Water demand in Lebanon is still undefined, because no creditable measures have been applied, and if these measures exist, they are found with obvious contradictory. In all cases, the largest part of the Lebanese territory is under water shortage and the demand for water is a national problem that remains unsolved since long time.

Added to water shortage and the intermittent water supply, there is also water pollution which becomes a widespread geo-environmental problem, and it reached both the surface and sub-surface water resources in the entire country. Hence, the current unfavorable situation on water resources in Lebanon creates conflicts between the supplier and the consumer. The inadequate supply, in some instances, results irresponsible and unethical behavior from consumers in different water uses, and thus consumers believe that they should manage their water needs and they cannot depend on the governmental sector to provide them with their water needs.

Recently, demand for water has been diverted by the consumers into chaotic exploitation manners, and they started utilizing water resources with no control and without any regulated methods. Therefore, direct water pumping from surface water sources (e.g. rivers, lakes, etc.) becomes a common phenomenon; even though small pools are dug near snowpack to collect the melting water and deliver it for long distances for agricultural and domestic purposes. In addition, groundwater exploitation has been chaotically increased and the number of private water wells are being in dramatic increase, notably in regions where groundwater is shallow and can be tapped easily with cheap cost such as the case in the Bekaa Plain where hundreds of water wells are dug every month. Most of these private wells have low discharge rate and they sometime become dry.

The recent estimates showed that there is a decline in the surface water resources between 55% and 60% over the last four decades (Shaban 2011). This also accompanied with an abrupt lowering (i.e. several tens of meters) in the water table for the major aquifers in Lebanon. The obvious decrease in the volume of surface and groundwater affected the nexus between water and the related sectors, thus it is reflected on the agriculture, energy, food security and it extended to the socioeconomic sector as well.

In Lebanon, the majority of water demand goes to irrigation purposes and many contradictory estimates have been illustrated for water allocation. For example: Jaber (1997) and Comair (1998) allocated water consumption in Lebanon as 12–32%, 8–18%, and 60–70% for domestic, industrial and irrigation purposes; respectively. Besides, World Bank (2003), illustrated 30.5%, 10.5% and 59% for domestic, industrial and irrigation; respectively.

The common estimates for irrigation range between 62% and 80%, and sometimes 85%.However, the recent applied survey showed that 72% is found the most convenient ratio even though this ratio is always oscillating from one year to another depending on the rainfall rate as well as on the demand for water for domestic and industrial purposes.

For the domestic and industrial sectors, they almost consume 26% and 6%; respectively (Shaban and Hamzé 2017). These ratio are also controlled by many factors (e.g. volume of water supply, partitioning period, etc.) and then they are found to be changed but with less percentage if compared with that of irrigation.

Water demand projections have been also applied to figure out future water requirements in Lebanon. According to SOER (2010), a projection for the years 2010–2035 have been done. It resulted that the demand for water in 2030 will be as: 583, 156 and 1050 million m^3 for domestic, industrial and irrigation uses; respectively (Fig. 1.2).

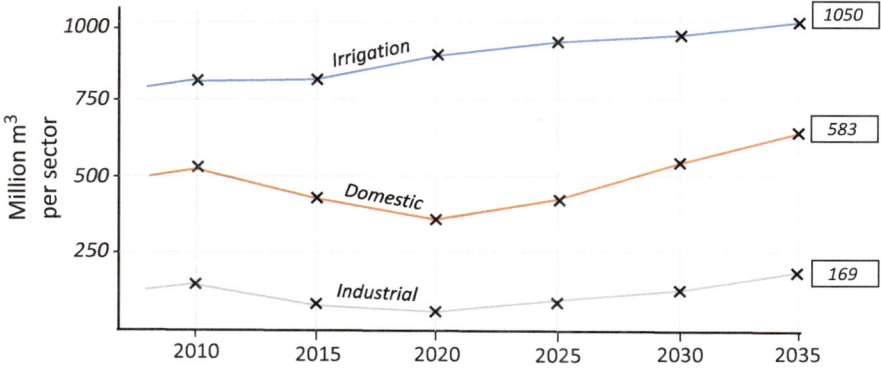

Fig. 1.2 Annual water demand by sector (2010–2035). (Adapted after SOER 2010)

Table 1.1 Water demand per capita in Lebanon as adopted from different sources

Year[a]	Consumption (m³/capita/year)	Source
1966	243	Jaber (1997), Comair (1998)
1966	236	Fawaz (2007)
1990	437	Jaber (1997), Comair (1998)
1990	297	Jaber (1997), Comair (1998)
1996	367	Fawaz (2007)
2000	592	El-Fadel et al. (2001)
2001	442	MoE (2011)
2010	307	MoEW (2010)
2011	220	Shaban (2011)
2015	526	Jaber (1997), Comair (1998)
2015	489	MoE (2011)

[a]Year of estimation

Water demand per capita in Lebanon has been recently estimated at about 217 m³/capita/year. In this regards, there are many estimates obtained by many researchers and the resulted values were totally different (Table 1.1). The following estimates show an example of the calculated water consumption per capita in Lebanon:

The discrepancy in water allocation also occurs between different regions in Lebanon. Therefore, it is always understood that water consumption in urban areas are greater than that in the rural ones. Nevertheless, recently the applied socioeconomic survey in selected regions revealed that rural areas demand amount of water larger than that in the urban ones. Therefore, the resulted survey showed that domestic water consumption are: 190 l/day/capita and 157 l/day/capita in the rural and urban areas; respectively.

1.6   Water Supply

Water supply is usually considered as the amount of water received from the public water sector, but when this amount becomes insufficient to cope with the consumer demand, alternative supply sources are adopted. In Lebanon, water supply has many aspects and it is dependent on miscellany of sources; however, it varies between rural, urban and densely urbanized rural regions (Fig. 1.3). The majority of water supply includes water from the obtained pipes by the public water sector, from groundwater boreholes, bottled water, water trading and harvested water (Shaban 2016).

Drilling water wells is a common phenomenon in Lebanon to compensate water shortage. These wells, which are mostly private ones, can contribute to a considerable range which is in average equal to 1/3 of water needs. These wells are widespread in the urban/or densely rural areas, such as in Beirut and Zahle (Fig. 1.3).

Bottled water, as an aspect of water supply, is only for domestic uses (e.g. drinking, cooking, etc.) and the highest percentage of bottled water distribution is in Beirut and its suburbs where sufficient/or pure water is lacking. The number of bottled water companies which are permitted by the Ministry of Public Health (MoPH) is 42 ones. They are 10, 15, 11 and 6 distributed in Mohafazat North, Mount-Lebanon, Bekaa and South; respectively (MoPH 2019).

Water trading has also a major contribution in water supply, and it is usually extracted either from boreholes in major cities like in Beirut area, or from the harvested water in the rural areas. In this respect, water trading is continuously increasing and different prices were noticed. However, the average price in Beirut rages between 100 and 125 LL/*l* and be between 50 and 75 LL/*l* in rural areas.

Stored water is another source of water supply in Lebanon. The largest part of this water is collected in wet season and then used in summer time and mainly for irrigation purposes. This aspect of water is usually harvested from rainfall or from the melting snow, notably in the mountainous rural areas where hill lakes and Earth reservoirs are established to store, in some instance, 1/3 of the water needs. However,

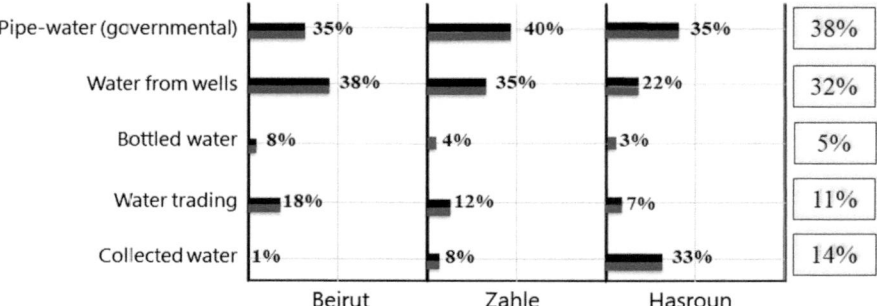

Fig. 1.3 Aspects of water supply in representative regions from Lebanon. (Adapted after Shaban 2016)

Surface water/Groundwater
supply percentage

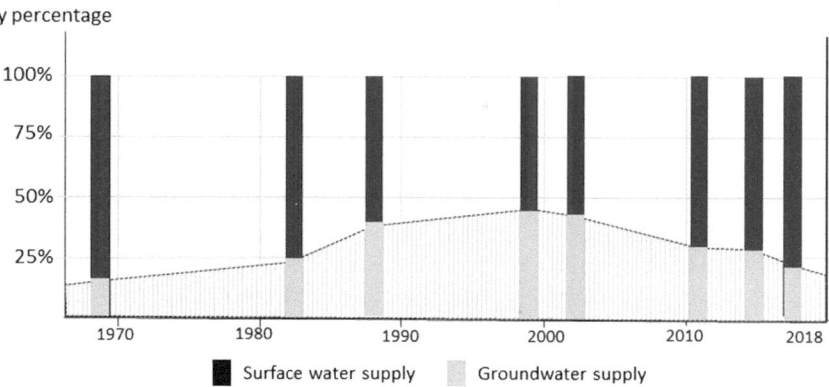

Fig. 1.4 Water supply from surface water and groundwater in Lebanon (1970–2018)

in urban areas like in Beirut, water collection is done in ponds and plastic tanks, and it is collected from the water supplied by pipes.

In generally, the current status in water supply is not satisfied. Thus, then water public sector is applying water partitioning, and it almost follows 2 days per week, and in many regions water sometimes do not reach consumers for a couple of weeks, notably in summer season or when maintenance is required for water networks. This indicates the poor supply and the urgent need for alternative management approaches.

It is clear that water supply form the public water sector does not exceed the 35% of the total water demand in all regions in Lebanon where the supply pipes are not well maintained and leakage from these pipes exceeds 40% in some cases. This unfavorable situation created new implementations; in particular, the increased number of illegal wells and water trading without quality controls.

In early 1970s, water from surface resources (notably from rivers) exceeded 85% of the total water supply in Lebanon, and the rest 15% was from groundwater pumping. Therefore, many changes have been developed and groundwater contribution reached about 45% of water supply in the year 2000. Lately, the adopted groundwater pumping regulations taken by the Ministry of Power and Water (MoPW) in Lebanon controlled the number of dug boreholes and then the pumping rate and this in turn created a new figure for water supply from surface water and groundwater resources (Fig. 1.4).

1.7   Previous Studies

Many studies on water resources in Lebanon have been produced, but they are mostly integrated with other themes (e.g. geology, agriculture, energy, political conflicts, etc.). From the geological point of view, they have been started since early 1830s when concerns on the geology of Lebanon has been raised. This was pioneered by the French geologists who worked in the region (Botta 1833; Blanche 1847). Further on, with the beginning of 1930s, another French geologist worked in the entire region of Lebanon and Syria; Louis Dubertret (*Father of Lebanon's geology*) started the geological mapping of the region at different scale including Lebanon (Dubertret 1933, 1953, 1955, 1963, 1966).

Since the early 1950s water studies were substantially raised in Lebanon. These studies belonged to different institutes concerned with water resources. Thus, the majority of these studies implies research papers, thesis dissertations, technical and assessment reports, projects, books. There are many of topics selected on these studies; in particular on hydrogeology such as:

(e.g. Yondanov 1962; water economics El-Qareh 1967; Majdalani 1977; Shaban 1987 and Khadra 2003); hydrology (e.g. Abd EL-Al 1953; Ghattas 1975; Daher 2015); karst hydrogeology (e.g. Hakim 1985; Edgell 1997 and Hamdan 2018); water quality (e.g. Abbud and Aker 1986; Peltikian 1980; Darwich et al. 2011); environmental hydrology (e.g. Khair et al. 1994; Jurdi et al. 2002; Fadel et al. 2015); water economics and policy (Fawaz 1969, 2007; Jaber 1995; Na'ameh 1995; Khawlie 2000b; CNRS-L 2015 and Shaban 2016); impact of climate change on water (e.g. Khawlie 2000a; Bou Zeid and El-Fadel 2002; Karam 2009; Shaban 2011 and Shaban et al. 2019); use of new techniques in water resources assessment (e.g. Shaban 2003 and Shaban et al. 2005).

Chapter Highlights
– Lebanon constitutes a climatic barrier that captures wet air masses blown by wind from the Mediterranean Sea.
– The entire Lebanese territory represents a water junction where three regional surface water systems exist.
– Water demand for is estimated at 1350 m^3/capita/year, and it has been decreased at by about 32% due to the geo-political conflicts and the displacement from neighboring regions.
– Water supply form the public water sector does not exceed the 35% of the total water demand.
– Except this book, there is no comprehensive study, including all aspects of water resources, has been elaborated for the entire Lebanon.

References

Abbud M, Aker N (1986) The study of the aquiferous formations of Lebanon through the chemistry of their typical springs. Lebanese Science Bulletin 2(2):5–22
Abd EL-Al I (1953) Statics and dynamics of water in the Syro-Lebanese limestone massif. Ankara symposium on arid zone hydrology, Ankara, Turkey, UNESCO, pp 60–76
Blanche C (1847) Lettre du Liban. Bull Soc Géol Fr., 2ème série, tome V:12–17
Botta P (1833) Observations sur le Liban et l'Anti-Liban. Mém Soc Géol Fr, t.I, mém:135–160
Bou Zeid E, El-Fadel (2002) Climate change and water resources in Lebanon and the Middle East. J Water Resour Plan Manage 128:5(343):343–355

CESFB (Center for Economic Studies at Fransabank) (2018) The policies and actions needed to face the growing water security challenges in Lebanon

CNRS-L (National Council for Scientific Research, Lebanon) (2015) Regional coordination on improved water resources management and capacity building. Regional project. GEF, WB

Comair FG (1998) Sources and uses of water from the Litani Basin and Karoun Lake. Paper presented at the Workshop on Pollution in the Litani Basin and Lake Karoun, and Environmental Problems in the Western Bekaa and Rashaya, 9–10 May 1998

Daher M (2015) Hydrologic assessment of Litani River basin in the view of climatic change. Unpublished M.Sc. Thesis, Lebanese University, Faculty of Agronomy, 34p

Darwich T, Atallah T, Francis R, Saab C, Jomaa I, Shaban A, Sakka H, Zdruli P (2011) Observations on soil and groundwater contamination with nitrate: a case study from Lebanon-East Mediterranean. Agric Water Manag 99:74–84

Dubertret L (1933) La carte géologique au millionème de la Syrie et du Liban. Rev Géo Phys Géol Dyn 6(4):269–316

Dubertret L (1953) Carte géologique de la Syrie et du Liban au 1/50000me. 21 feuilles avec notices explicatrices. Ministère des Travaux Publics. L'imprimerie, Catholique, Beyrouth, 66p

Dubertret L (1955) Carte géologique de la Syrie et du Liban au 1/200000me. 21 feuilles avec notices explicatrices. Ministère des Travaux Publics. L'imprimerie, Catholique, Beyrouth, 74p

Dubertret L (1963) Liban et Syrie. In Lexique Stratiraphique International. Asie. CNRS, France, 10, 11–153

Dubertret L (1966) Liban, Syrie et bordure des pays voisins: Notes Mem. Moyen Orient 8:251–358

Edgell H (1997) Karst and hydrogeology of Lebanon. Carbonates & Evaporites 12:220–235

El-Fadel, M, Zeinati M, Jamali D (2001) Water resources managementin Lebanon: institutional capacity and policy options. Water Policy 3:425–448

El-Qareh R (1967) The submarine springs of Chekka: exploitation of a confined aquifer discharging in the sea. Unpublished M.Sc. thesis, American University of Beirut, Geology Department, 80p

Fadel A, Atoui A, Lemaire BJ, Vinçon-Leite B, Slim K (2015) Environmental factors associated with phytoplankton succession in a Mediterranean reservoir with a highly fluctuating water level. Environ Monit Assess 187(10):633

Fawaz M (1969) Water policy of Lebanon. Beirut Club, Nov. 22 (in Arabic)

Fawaz M (2007) Towards a water policy in Lebanon. Published book, ESIB (In Arabic), Beirut,Lebanon, 259p

Ghattas I (1975) The geology and hydrology of the western flexure of Mount Lebanon between Dbaiyeh and Jdaide. Unpublished M.Sc. thesis, American University of Beirut, Geology Department, 99p

Ghiotti S, Riachi R (2013) Water Management in Lebanon: a confiscated reform?, Etudes rurales: 2013/2 (no 192). Editions de l'E.H.E.S.S.135–152

Hakim B (1985) Recherches hydrologiques et hydrochimiques sur quelques karsts méditerranéens: Liban, Syrie et Maroc. Publications de l'Université Libanaise. Section des études géographiques, tome II, 701p

Hamdan A (2018) Estimation of transit times in karst aquifers using environmental tracers (tritium, helium, chlorofluorocarbons, and sulfur hexafluoride) : application on the Jeita aquifer system-Lebanon. Unpublished M.Sc. Thesis, American University of Beirut, Geology Department, 147p

Jaber B (1995) Water problems of Lebanon. National Congress on Water Strategic Studies Center. Beirut (in Arabic), 67p

Jaber B (1997, April) Water in Lebanon: problems and solutions Public lecture given in the Department of Hydrology. Purdue University, Lafeyette, IN, USA

Jurdi M, Korfali S, Karahagopian Y, Davies B (2002) Evaluation of water quality of the Qaraaoun Reservoir, Lebanon: suitability for multipurpose usage. Environ Monit Assess 77(1):11–30

Karam F (2009) Climate change and variability in Lebanon: impact on land use and sustainable agriculture development. Unpublished report, available on http://www.fao.org/sd/climagrimed/pdf/ws01_24.pdf

Khadra W (2003) Hydrogeology of the Damour Upper Sannine-Maameltain aquifer. Unpublished M.Sc. thesis, American University of Beirut, Geology Department, 235p

Khair K, Aker N, Haddad F, Jurdi M, Hachach A (1994) The environmental impact of human on groundwater in Lebanon. Water, air and pollution, vol 78. Kluwer Academic Publications, pp 37–49

Khawlie M (2000a) Assessing water resources of Lebanon in view of climate change. Workshop on: soil and groundwater vulnerability to contamination. ACSAD-BGR, Beirut, 7–10/2/2000, 19p

Khawlie M (2000b) Environmental problems related to water resources in Lebanon and requirements for sustainability. Workshop on: Integrated Management & Sustainable Use of Groundwater & Soil Resources in the Arab World. ACSAD

Lonergan S, Brooks D (1994) Watersheds: the roles of fresh water in the Israeli-Palestinian conflict. International Development Research Center. IDRC Books

Maïla-Ateihe A (2009) Le Site de Nahr El-Kaleb.BAAL. Bulletin d'archeologie et d'architecture Libanises. Hors-serie V. Ministry of Culture, 344p

Majdalani M (1977). Geology and hydrogeology of the Faraya-Afqa area, central Lebanon. Unpublished M.Sc. thesis, American University of Beirut, Geology Department, 143p

Mallat H (2003) Droit de l'urbanisme, de la construction, de l'environnement, et de l'eau au Liban. Bruylant, Brussels

MoE (Ministry of Environment) (2011) Lebanon's second National Communication to the United Nations framework convention on climate change. Accessed online in November 2013 at: http://www.moe.gov.lb/ClimateChange/snc.htm

MoEW (Ministry of Energy and Water) (2010) National Water sector strategy. Available at http://climatechange.moe.gov.lb/viewfile.aspx?id=183

MoPH (Ministry of Public Health) (2019) List of bottled water permitted by MoPH. Avalable at:

Na'ameh M (1995) Water problems of Lebanon. National Congress on Water Strategic Studies Center. Beirut (in Arabic), 67p

OEB (Office des Eaux de Beyrouth) (1996) Eaux de Beyrouth: Centenaire de l'usine de Dbayeh, Beyrouth, p 14

Peltikian A (1980) Groundwater quality of greater Beirut in relation to geological structure and the extent of seawater intrusions. Unpublished M.Sc. thesis, American University of Beirut, Geology Department, 107p

Shaban A (1987) Geology and hydrogeology of the Nabatieh area. Unpublished M.Sc. thesis, American University of Beirut, Geology Department, 103p

Shaban A (2003) Etude de l'hydrogéologie au Liban Occidental: Utilisation de la télédétection. Ph.D. dissertation. Bordeaux 1 Université, 202p

Shaban A (2011) Analyzing climatic and hydrologic trends in Lebanon. J Environ Sci Eng 5(3)

Shaban A (2016) New economic policies: instruments for water management in Lebanon. Hydrol Curr Res 2016(7:1):1–7

Shaban A, Hamzé M (2017) Shared water resources of Lebanon. Nova Publishing, New York, p 150

Shaban A, Houhou R (2015) Drought or humidity oscillations? The case of coastal zone of Lebanon. J Hydrol 529(2015):1768–1775

Shaban A, Khawlie M, Abdallah C, Faour G (2005) Geologic controls of submarine groundwater discharge: application of remote sensing to North Lebanon. Environ Geol 47(4):512–522

Shaban A, Awad M, Ghandour A, Telesca L (2019) A 32-year aridity analysis: a tool for better understanding on water resources management in Lebanon. Acta Geophys. https://doi.org/10.1007/s11600-019-00300-7

Shahin, M (1996) Hydrology and Scarcity of Water Resources in the Arab Region. Balkema/Rotterdam/Brookfield, 137p

SOER (2010) The State and Trends of the Lebanese Environment. Ministry of Environment, UNDP. 355pp

UNDP, FAO (1983) Spate irrigation. Prove sub-regional experts consultation on Wadi development for agriculture. In AG: UNDP/RAB/84/030

World Bank (2003) Republic of Lebanon – Policy note on irrigation sector sustainability. Rep. No. 28766 – LE

Yondanov Y (1962) Aperçu succinct sur l'hydrogéologie du Liban. Land and Water Development Co., Unpublished Report, 52p

Chapter 2
Atmospheric Regime and Terrain Characteristics

Abstract Usually the regime of the atmosphere, including mainly the precipitation and temperature, controls the volume of water. While the characteristics of terrain, including surface and sub-surface properties govern water flow and storage. These two physical pillars must be primarily identified in order to reach optimal water resources assessment. Hence, the relatively humid climate is interlinked with the rugged topography and both govern water discharge, including water flow energy, accumulation, and infiltration, storage in the substratum and even in water loss to the sea. Therefore, the atmospheric regime and the characteristics of a terrain are principal generators for the entire water cycle like the case in Lebanon. The climate of Lebanon is known by wet periods that are pronounced by the rainfall and the snow cover for a considerable number of months over the year. Besides, the terrain has different responding features. This implies accelerating surface water flow and regular water infiltration among the rock masses, as well as the chaotic groundwater flow into the karstic conduits. This chapter illustrates the major atmospheric variables and the terrain characteristics-induced water resources for the entire Lebanon.

Keywords Rainfall rate · Geomorphology · Snow cover · Water flow · Lithological characteristics

2.1 Atmospheric Variables

There are two principal atmospheric variables usually considered while applying climatic assessment and projections. These variables are the precipitation and temperature, which can fulfill the scope of climatic trend analysis. While, using additional variables (e.g. humidity, sunlight radiation, wind, etc.) would be much more favorable. However, this is often dependent on climatic data availability whether from ground measures or from space techniques.

© The Editor(s) (if applicable) and The Author(s), under exclusive license to
Springer Nature Switzerland AG 2020
A. Shaban, *Water Resources of Lebanon*, World Water Resources 7,
https://doi.org/10.1007/978-3-030-48717-1_2

In Lebanon climatic data are almost available; however, complete time series usually do not exist, and sometimes data are available only for specific regions where meteorological stations are functional. Yet, there is always a debate on measuring climatic variables in Lebanon and this can be attributed to the following:

1. The non-uniformity of Lebanon's terrain, notably the rugged topography which needs increased number of meteorological stations to have typical representation of climate measures.
2. Data reliability notably that many meteorological stations need maintenance and some of them are too old.
3. The lack to the exchange and coordination between different institutions who own climatic data series.

Due to its rugged topography, Lebanon has diverse climatic regime whereas the precipitation is unevenly distributed, and then resulting microclimatic zones. This is well pronounced because it winds from several directions and with warm, cold and humid weathers (Arkadan 2008). In addition, there is meteorological turbulences in the blowing air, and thus, rain shadow impact occurs among the topography of the Lebanese mountains. Therefore, cold air masses rising up on the mountains from west and descending air warms and picks up available moisture to the east of the mountain ridges resulting eventually climatic turbulences (Shaban and Houhou 2015).

There are many descriptions illustrated for the climate of Lebanon, while it is commonly mentioned that Lebanon is a Middle East Region with *semi-arid* climate without any evidenced indicator for such understanding. However, FAO (2009) stated that the Lebanon's climate is typically Mediterranean climate where heavy rains in winter (November to May) and dry and arid conditions in the remaining months of the year. Meanwhile, Lebanon is describing as semi-arid area during the dry season to humid to sub-humid in the wet season (NC 2011). While CAS (2015) stated that Lebanon is attributed to the northern temperate zone. Besides, CAL (1982) pointed out that Lebanon has an oceanic climate during winter and a subtropical climate during summer.

Lebanon is always attributed to a geographic area among wide regions (e.g. MENA, Arab, ME, ESCWA, etc.) when applying regional climatic trends, projections and scenarios. However, this is merely erroneous, because Lebanon has a remarkable mountainous geography, which is totally distinguished from the surrounding areas.

In fact, the territory of Lebanon is characterized by typical Mediterranean climate, which is moderately cold and rainy in winter, warm in summer, mild in spring and autumn (Shaban 2003). In this respect, the available climatic records are generally insufficient to build creditable assessment for the atmospheric condition unless advanced statistical methods (e.g. Singular Spectral Analysis, Fisher-Shannon, etc.) are applied (Lovallo et al. 2013; Telesca et al. 2013).

2.1.1 Sources of Climatic Data

Climatic data in Lebanon are derived from ground meteorological stations, which are fixed in several localities in the country. The largest number of these stations are mandated by the Ministry of Public Works and Transport (MoPWT) where it provides meteorological data through the Climatic Atlas of Lebanon-CAL (1971, 1973 and 1982) and General Directorate of Civil Aviation GDCA (1999).

Until the early 1990s, there were about 70 meteorological stations in Lebanon where about 50 of them were in the western part of the country, and more specifically in the coastal zone and the adjacent occidental mountain. However, most of these stations were not functional (i.e. lack for maintenance, political conflict, etc.) except- 20–25 ones. Recently, there are about 125 operational meteorological stations in Lebanon fixed/or operated by different institutions, such as: Lebanese Agriculture Research Institute, Lebanon (LARI 2017) who has 60 operational stations; while some data are delivered by researchers such as Ghaddar (1995) and Na'ameh (1995).

Precipitation data were also extracted from remotely sensed products and more certainly from the Tropical Rainfall Mapping Mission (TRMM) which was extended by NASA since 1998 (2015), Climate Hazards group Infrared Precipitation with Stations (CHIRPS 2017); and from NOAA climatic data system – National Oceanographic Data Center (NOAA 2013).

2.1.2 Precipitation

In general, precipitation includes rainfall and snow, and it is considered as the major source of water in Lebanon where it makes it a country with water renewability. However, there obvious diversity in precipitation rate as a result of the rugged topography and then the altitude-controlled climate. Hence, rainfall rate ranges between 650 mm and 1500 mm as it distributed between the coastal zone and the crests; respectively. Therefore, abrupt changes in the rainfall rate often occurs between relatively short land distances (Fig. 2.1).

Solid precipitation (snow) is another aspect of precipitation in Lebanon where it is considered as a principal water-feeding source for surface water and groundwater. Thus, Lebanon is the only geographic region in the Middle East where snow cover remains for few months on the mountains and covers about 2250–2500 km^2.

The precise average rate of rainfall in Lebanon remains uncalculated till few years ago. Thus, several contradictory estimates have been mentioned. These can be (as examples) illustrated as: 982 mm, 940 mm, 820 mm, 873 mm, 825 mm, 823 mm, and 800 mm according to Rey (1954), UNDP (1970), Plassard (1971), CAL (1982), Geadah (2002), WB (2003), MoEW and UNPD (2016), respectively.

According to the measured data series on rainfall in Lebanon between 1950s and 2015 (CNRS-L 2015), there are rainfall peaks in Lebanon often between November

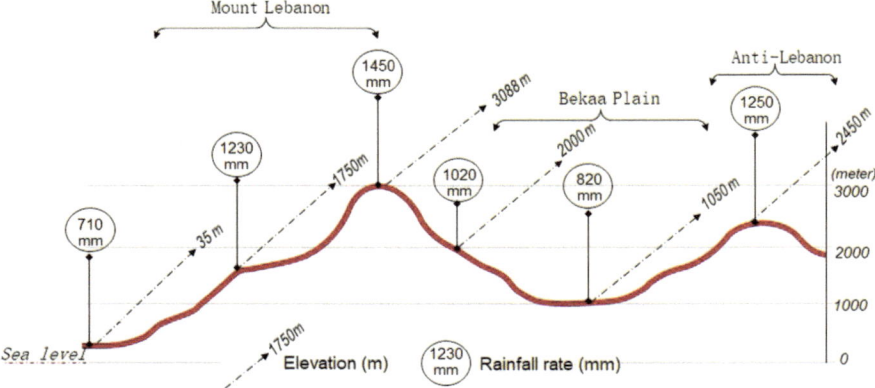

Fig. 2.1 Schematic representation for the distribution of rainfall rate long different Lebanese lands

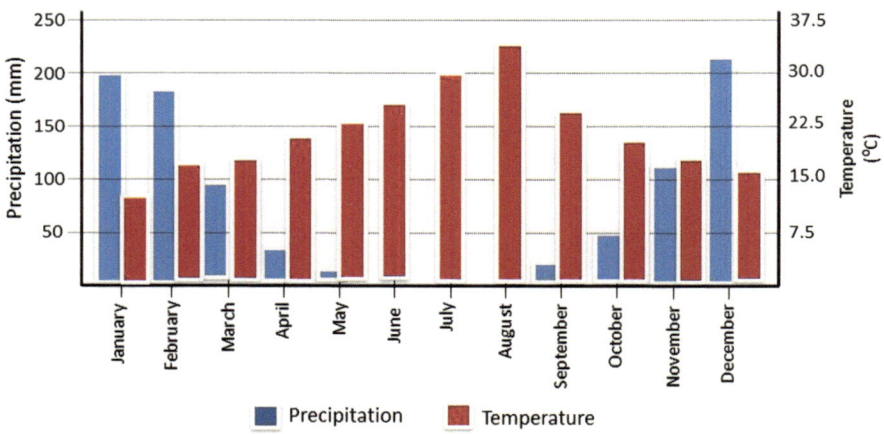

Fig. 2.2 Average annual rainfall and temperature rate

and February where the average monthly rate reaches to about 200 mm, while the lowest rates of rainfall exist between May and August months (Fig. 2.2).

Shaban and Houhou (2015) calculated the average rainfall rate in Lebanon over more than six decades (i.e. since 1950s). They used records from different ground stations and also from remotely sensed products covering diverse zones from Lebanon. The collected records were sorted and interpolation approaches were applied where gaps exited. Thus, the resultant shows that the average rainfall in Lebanon is 910 mm. While the estimated raining days ranges between 60 and 70 (± 15) days per year mainly between October and April. When integrating the topography of Lebanon with the distribution of rainfall, it was obvious that six climatic zone are resulted (Table 2.1).

Table 2.1 The major climatic zones of Lebanon

Climatic zone	Altitude range (m)	Rainfall rate (mm)	Main geography[a]
The Coastal ribbon	<100	550–850	Coastal plain (<5 km wide)
The Plateau	100–1100	750–1050	Diverse, moderate heights
Western Heights	>1100–3088	1000–1550	The maximum at Kornet Es-Sawda of 3088 m (Mount-Lebanon)
Bekaa Plain North	900–1200	200–450	An open land between two mountain chains
Bekaa Plain South	800–1200	600–1100	An open land where several hills occur
Eastern Heights	>1200	700–1250	The maximum at Tallet Mousa at 2620 m (Anti-Lebanon)

[a]For the entire Lebanon

2.1.3 Temperature

Even though, the measuring of temperature is simple enough to apply and it is usually the primary measure fixed on meteorological stations, yet data on temperature in Lebanon is also lacking. This is because of the rapid changing of temperature rate with time.

The temperature rate has been obviously increased in Lebanon and it affected several sectors, notably the agriculture one. However, the temperature of Lebanon is considered as moderate and cannot be described as hot or very cold even during the extreme weather. It is merely an altitude-related climatic variable, and that is why it is abruptly changing in space and time.

There are many contradictory estimates for temperature rates in Lebanon. Hence, the average annual temperature ranges between 14 °C in winter to 27 °C in summer in the coastal zone, and thus averaging 21 °C. While, in the mountainous areas, the average annual temperature is below 12 °C, ranging from 2 to18 °C in summer in winter; respectively.

According to the analyzed time series applied between 1950 and 2015 by CNRS-L (2015), the hottest temperature rates in Lebanon were found between June and August, while the coldest months are December and January and sometimes extending to February (Fig. 2.2).

Lately, temperature trends have been analyzed using long time series for the entire Lebanon after collecting all the available records (Shaban and Hamzé 2017). Hence, the resultant shows that the lower calculated average temperature ranges between 5 °C and 10 °C in regions above altitude of 1800 m; around15° C on areas higher than 1100 m; and it is about 20 ° C in the coastal zone. Therefore, the distribution of average annual temperatures with different altitudes can be summarized as follows:

- Coastal zone (<250 m) = 24 °C
- Plateau (250–750) = 22.5 °C
- Moderately elevated areas = (750–1500) = 19.5 °C
- Heights (1500–2500) = 18.5 °C
- Crests (>2500) = 14 °C.

2.1.4 Evapotranspiration and Relative Humidity

Measures of evapotranspiration (Et) in Lebanon are still inadequate, notably that the number of fixed lysimeters are few enough to establish a comprehensive estimation for Et. However, many estimations were put, as separate measures where almost all of these estimations adopted that Et in Lebanon ranges between 48% and 53% of the precipitated water volume. Thus 48%, 50%, 52% and 53% were put by Fawaz (1992); MoE/ECODIT (2002); MoEW (2010) and Shaban (2003); respectively.

According to FAO (2009), the mean annual potential evapotranspiration ranges between 1100 mm in the coastal zone and 1200 mm in the inner zone (the Bekaa Plain). While Karam et al. (2003) calculated the annual potential evapotranspiration at 936 mm for the entire Lebanon.

Lately, the advanced Et measuring technique has been involved in several studies applied in different regions from Lebanon, especially in the Bekaa Region where agriculture is most dominant (Awad 2019). Thus, two Bowen ratio stations have been installed in the Bekaa Region by the CNRS-L to measure crop yield and Et (Fig. 2.3).

With respect to the relative humidity, as a function of temperature, it is usually measured when the functional meteorological stations are found. Generally, the relative humidity in Lebanon is constant along the coastal zone where it is about 70%, and decreases in the Southern regions to about 65% (Farhat 2018). While, in mountainous regions the relative humidity is often oscillating between 60% and 80% in winter; besides 40 and 60% in summer.

2.1.5 Aridity

In contrast with other measurable meteorological variables, aridity cannot be measured directly by the meteorological station. Indeed, it is a function of rainfall and temperature and it is simply identified as a long-term lack of rainfall/or moisture and it is a result of the presence of dry, descending air.

Many formula are applied to calculate the aridity index, such as: De Martonne, Thornthwaite, Koeppen, etc. The Emberger aridity index (E_{Ai}), as obtained by Emberger (1932) is much reliable due to the fact that it integrates two climatic variables, the mean annual rainfall (R) and the mean temperature (T) where the latter includes the hottest and coolest temperatures according to the following formula.

Fig. 2.3 Bowen ratio
station (NESA) installed in
the potato field

$$EAi = \left[(T-t) \times \frac{\dfrac{1000 \times R}{(t+T)}}{2} \right]$$

Due to its significance in characterizing the climate of Lebanon, aridity over more than 30 years was recently calculated by the author and other (Shaban et al. 2019). The calculated Emberger aridity index was obtained using detailed inventory for the long-term analysis. In this view, remotely sensed products were also integrated to provide measures on rainfall and temperature for ten investigated stations located in diverse topography in Lebanon.

As a result, Lebanon's climate can be described as sub-humid. In particular, Lebanon would not be remarkably affected by the drought which is the common phenomenon for the entire Middle East Region (Shaban et al. 2019).

Another major outcome on the aridity index in Lebanon is the periodicity (i.e. recurrence of aridity periods), which was identified using a robust technique. Therefore, the periodicity of aridity in Lebanon is altitude related and it ranges between 2 and 21 years (Telesca et al. 2018). Two major geographic zones represent the geographic distribution of periodicities in Lebanon; these are in the northern and southern parts. Therefore, the remarkable diversity in periodicity evidences obvious microclimates with specified weather cyclonic climate that characterizes the climate of Lebanon rather than a merely existence of climate change (Telesca et al. 2018).

2.2   Morphology

Among the Middle East Region, Lebanon can be well noticed with its unique topography where elongated mountain chains are laying parallel to the Mediterranean Sea. These ridges, along the Eastern Mediterranean Basin, represent a climatic fence which makes Lebanon with unique climate if compared with the surroundings.

The majority of topography of the Lebanese territory is composed of three main morphological units. These are: (1) uplifted mountain chains entitled Mount-Lebanon which are located adjacent to coast, (2) compressed mountain chains tilted eastward and described as Anti-Lebanon and (3) a wide depression called the Bekaa Plain, which extends in between the two mountain chains. Therefore, these units are directed in the NNE-SSW direction (Fig. 2.4). These units can be described as follows:

1. Mount-Lebanon: This is the western mountain chains of Lebanon which are extending parallel to Mediterranean Sea. From these chains, the entire shoreline can be observed. Mount-Lebanon includes the coastal plain (a ribbon with less than 5 km wide), moderately elevated zones (plateau up about 1000 m altitude) and the elevated region of western mountains with an average altitude of about 1400 m and then ascending up to 3088 m at Kornet Esawda, the highest crest in the Middle East Region.

    Shaban (2003) classified the topographic units of Mount-Lebanon as the: Coastal Plateau (<100 m), Lower Slopes Plateau (<500 m), Upper Slopes Plateau (500–1500 m) and Elevated Crests (>1500 m).

    Mount-Lebanon encompasses acute surfaces with sharp relieves where consequent, subsequent and insequent valleys and gorges are dominant. While, the average slope gradient ranges between 60 and 140 m/km and averaging about 80 m/km.

2. Anti-Lebanon: This represents the mountain chains to the east and they shape the border with Syria. They differ from the chains of Mount-Lebanon in that they are relatively with smooth surface relief since they are resulted of folded carbonate rocks. Thus, Anti-Lebanon ranges has an average altitude of about 1050 m. While the average slope gradient ranges between about 50 and 90 m/km, and thus averaging about 65 m/km.

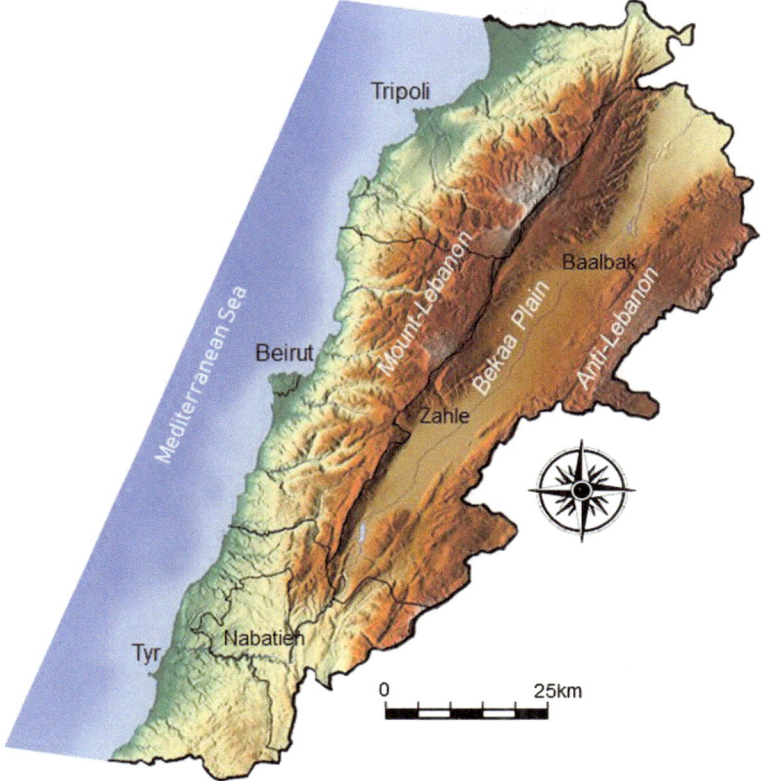

Fig. 2.4 Major morphological units of Lebanon

3. Bekaa Plain: It is a depression-like shape that situated between the previous two
   geomorphologic units (Fig. 2.4). Even though it had two major slope directions
   (i.e. to the north and the south), yet the average altitude of the Bekaa Plain is
   about 850 m, while it has a varied width that ranges between 8- and 10 km and
   get narrow in the southern side to reach less than 1 km. besides, the slope gradient does not exceed 10 m/km.

2.3   Karst Topography

Between 70% and 75% of the Lebanese territory is composed of exposed carbonate
rocks where limestone, dolomitic limestone and dolomite are widespread on terrain
surfaces. More than 85% of these carbonate rocks are karistified and the dissolution
of carbonates is well pronounced with a variety of aspects including surface and

sub-surface rocks. The karstification resulted irregular and rugged topography that can be well recognized even on the topographic maps were non-uniform contour alignments occur.

There are many studies done of the karst of Lebanon including mainly karst hydrogeology and its relationship to water flow and storage. Thus, understanding this geological phenomenon and its mechanism is essential, notably in groundwater assessment approaches (e.g. Guerre 1969; Edgell 1997; Hakim 1985; Khawlie and Shaban 1998; Azar 2000 and Shaban et al. 2000). In addition, karst was also studied in terms of geomorphology, landscape and surficial processes (e.g. Hakim 1985; Sanlaville 1977; Kheir et al. 2001).

Karstification in Lebanon occupies the surface and sub-surface rocks as well as it represents the contact surface between both where different scales and aspects occur as follows:

1. Surface karstification:

There is a wide variety of surface karstification in Lebanon which is often attributed to specific altitude and rock lithology. Yet, there is no define map showing the geographic distribution of different aspects of surface karstification. However, the most known aspects are: (1) sculptured and etched rocks (e.g. pinnacle karren, lapis, iceberg karren, rinnenkarst, spitzkarst and rill karst, (2) pavement karst (e.g. irregular and smooth surfaces), (3) cockpit karst (e.g. pits, dots and grooves) (4) channel and holes (e.g. cylindrical holes, solution pan tunnel, furrows).

2. Contact-surface karstification:

At arid climatic conditions where carbonate rocks are exposed; however, the dissolution processes on terrain surfaces work at high rate, and then geomorphologic linkages between surface and sub-surface karts occur. This in turn increases the rate of surface water percolation into substratum along the existed fissure and joints. Among this type of karst topography, sinkholes (i.e. dolines) are well known in Lebanon. These geomorphologic features often exist in altitudes above 2100 m (Shaban and Khawlie 2006).

The dominant sinkholes in Lebanon are found on the highest crests (Fig. 2.5). Good exposures are located in Ayoun Asseman, Jabal Hermoun and Kornet Es-Sawda. These sinkholes have a variety of dimensions that start from 1 m × 0.25 m to 300 m × 5 m in diameter and depth; respectively. In addition, caves, cavities and shafts are another aspects of contact-surface karstification in Lebanon which possess the geomorphological and hydrogeological characteristic of both the surface and sub-surface karstification.

3. Sub-surface karstification:

From the dimensional point of view, the sub-surface karstification is much more developed than those on terrain surfaces, notably that the underground carbonate dissolution is often characterized by rapid etching rate of the carbonate lithologies due to the geochemical conditions available, notably the existence of CO_2 and

Fig. 2.5 Sinkholes, a dominant contact-surface karst in Lebanon

sustainability of underground water seeps. This is also well pronounced where surfaces of weakness occur, such as faults, fissures and joints.

The shape, intensity and density of sub-surface fissures control the direction and development of the future karstification. These fissures are either structural/ lithological/or climatic features, since they reflect controls of all (Khawlie and Shaban 1998).

The majority of sub-surface karstification in Lebanon implies tremendous conduits and underground tunnels, grottos and galleries which are, in many instances, connected with the contact-surface karstification. Nevertheless, the sub-surface karstification is considered as a major groundwater-bearing reservoirs, such as those grottos of Jietta and Kottin Kafer Heem grottos. Moreover, most of these aspects of sub-surface karstification seeping groundwater on terrain surfaces and form most of the major issuing springs in Lebanon (e.g. Jeitta Spring, Ain Ed-Deleb Spring, etc.).

In Mount- Lebanon, more than 300 major caves have been identified, where 35% of them are found in the Jurassic dolomite rocks, while the rest occur in the Cenomanian dolomitic limestone.

There are many classifications done to describe the geographic distribution of karst topography in Lebanon and their physical characteristics. However, the classification plotted by Khawlie and Shaban (1998) was comprehensive (Table 2.2). Even though, this classification was attributed to the Mount-Lebanon; however, it can represent the entire Lebanese territory.

Table 2.2 Types of karst topography in Lebanon

Physical characteristics			Low altitude	Moderate altitude	High altitude
Physiography	Altitude (m)		< 500	500–1000	> 1000
	Climate	PPt	850 mm	Up to 1000 up to mm	Up to 1400 mm
		Sf	< 5 days/year	20–25 days/year	70–80 days/year
	Roughness		Smooth	Irregular	Slightly irregular
Geologic formation (major)			Cenomanian, Neogene; dolomitic limestone	Jurassic: dolomite and dolomitic limestone	Cenomanian: dolomitic limestone
Dominant Karstic features	Aspects		Small-scale features covered by soils (e.g. rillenkarst, cylindrical holes, furrows and pits)	Karst with different dimensions (sculptured linear rocks such as lapis, rinnenkarst, spitzkarst and caves	Karst with different dimensions. Solution depressions as sinkholes and dolines are dominant. Caves occur at large depth.
	Dimensions (major types)		Less than 1 m for the three dimensions	Up to 10 m height, 0.5 m diameter for the surficial karst. Several kilometers areal extent for the sub-surface ones	Up to 10 m depth, 5 m diameters for the depressions. Large extent for the caves.
	Abundance		Abundant on a wide range	Scattered in different mountainous localities	Dominant on the flat surfaces of the crests
Major controls			Geomorphology	Lithology	Structure
			Climatic		

Adapted after Khawlie and Shaban (1998)
PPt: Precipitation
Sf: Snowfall

In Lebanon, there are no estimations on the dissolution rate of carbonates. Nevertheless, Hakim (1985) relatively described the rate of dissolution through different exposed karstic rocks as follows:

- Cenomanian rocks at low altitudes = 15–25 $m^3/km^2/year$.
- Cenomanian rocks at moderate altitudes = 25–40 $m^3/km^2/year$.
- Jurassic rocks at high altitudes = 40–60 $m^3/km^2/year$.

2.4 Geology

With its small area, the geology Lebanon has been described in several studies done by L. Dubertret (1933) put the first geological elements for the entire Lebanon. The exposed rocks in Lebanon are relatively few in terms of the lithological characteristics where sedimentary rocks essentially compose the main rock sequence plus some scattered igneous rock bodies. Yet, the structural geology is not well understood due to its complexity, notably that Lebanon is a part of the Red Sea Rift System along the eastern Mediterranean Basin and constitutes a part of the unstable shelf of the Middle East. Therefore, tremendous rock deformations resulted, and hence they were non-uniformly merged the rock stratum.

According to Beydoun (1988), the exposed rocks in Lebanon are significantly influenced by the Upper Eocene and Oligocene Alpine Orogeny which is accompanied with complicated tectonic systems and then produced several folding and faulting processes. Beydoun (1972), described the mountain chains of Lebanon as uplifted blocks that make bended mountain chains. Thus, the uplifted blocks comprise horst anticline where flexturing and normal faulting shaping it flanks (i.e. along Mount-Lebanon and Anti-Lebanon). While the Bekaa Plain forms a regional graben syncline that marked by the Yammounah Fault from the west.

2.4.1   Lithostratigraphy

The oldest exposed rocks in Lebanon, are attributed to the Middle Jurassic age where marine environment deposition continued to the Middle Eocene age. Therefore, the huge bodies of the carbonate rocks composed the major rock succession, where 65–70% of Lebanon's territory exposes limestone, dolomitic limestone and dolomite. Intervening with the carbonate rocks; however, marl, sandstone and argillaceous rocks are found among the rock sequence. In addition, igneous rocks and more certainly the volcanic ones are exposed as basalts and tuffaceous rocks.

As reported by many studies, with a major focus on studies obtained by Dubertret (1953, 1955) and Beydoun (1972, 1977 and 1988), the lithostratigarphic sequence of Lebanon is composed of 17 major rock formations where several diverse rock units are interbedded among this sequence. The lithostratigarphic characteristics of these rock formations can be summarized in Table 2.3 and discussed in details as follows (Shaban 2003):

1. Jurassic Period: It represents the core of the exposed rock masses in Lebanon where its geographic distribution is merely structurally controlled by the presence of folding structures and sets of fault systems. The rock formations of the Jurassic Period are dominant in the Middle part of Lebanon (Keserwan Region) and in the south-east part on Hermoun Mountain, as well as a large patch occurs in the north.

– Middle Jurassic Epoch: It comprises the (1) *Callovian* (650 m: massive and thick bedded rocks, fractured and karistified dolomite and dolomitic limestone with some marl intervals); and the (2) *Oxfordian* (variable thickness, volcanic tuffs and rock mixtures that intervening with the deposition of detrital and oolithic limestone, marls and shale).

Table 2.3 Stratigraphic sequence of rock lithologies in Lebanon

Era	Period	Epoch	Age	Lithology	Symbol	Thickness[a] (m)	Area[b] (km^2)
Cenozoic	Quaternary	Holocene		Marine deposits, river terraces	Q	Variable	1466
		Pleistocene		Dunes, alluvial deposits			
	Tertiary	Pliocene		Limestone, marl, volcanics	P	360	269
		Miocene	Vindobanian	Conglomeratic limestone	m_2-ncg	320	183
			Burdigalian	Marly limestone	m_1	80	496
		Eocene	Lutetian	Limestone, marly and chalky limestone	e_2	800	786
			Ypresian	Chalky limestone, marly limestone and marl	e_1	370	271
Mesozoic	Cretaceous	Upper	Senonian	Chalky marl	C_6	400	742
			Turonian	Marly limestone, marl	C_5	200	297
		Middle	Cenomanian	Limestone, dolomitic limestone, marl	C_4	700	4125
			Albian	Marly limestone, marl	C_3	200	309
			Aptian	Dolomitic limestone	C_{2b}	50	184
				Argillaceous sandstone, marl, limestone	C_{2a}	250	208
			Neocomian-Barremian	Quartzite sandstone, mixed with clayey	C_1	Variable	436
	Jurassic	Upper	Portlandian	Oolitic limestone	J_7	180	139
			Kimmeridjian	Dolomite, limestone	J_6	200	1507
		Middle	Oxfordian	Volcanic materials, marly limestone	J_5	Variable	154
			Callovian	Dolomitic limestone, limestone	J_4	650	346

Note: Cretaceous basalt was noted mentioned since it is not often exist
[a]Cumulative thickness (exposed and hidden)
[b]Exposed area within Lebanon

– Upper Jurassic Epoch: It has two rock formations, the (1) *Kimmeridjian* (200 m: massive and thick bedded, highly fissured, jointed and karistified limestone and dolomitic limestone, interbedded with thin marly limestone horizons, in addition to tremendous chert nodules; and the (2) *Portlandian* (180 m: oolithic limestone and marly limestone intercalations. Nevertheless, this rock formation does not exist in many geographic as an evidence of non-sedimentation or differential erosion occurred during the Late Jurassic uplift).

2. Cretaceous Period: Cretaceous rock formations in Lebanon occupy the largest area of rock exposures. Similarly to the Jurassic rock, the geographic distribution of Cretaceous rocks is also structurally controlled. Thus, they are well observed in the coastal zone including the plateau and elevated regions as well as in the Anti-Lebanon ranges.

– Lower Cretaceous epoch: This epoch includes three rock formations in Lebanon. These are the (1) *Neocomian-Barremian* (variable thickness, calcareous sandstone with intercalations of siltstone, lignitic clays, interbedded with shale and sandy limestone, in addition to basaltic intrusions and tuff making the boundary between Jurassic and Cretaceous); (2) *Lower Aptian* (250 m:moderately thick-bedded of clastic limestone interbedded with marl and argillaceous and sandy limestone and shale); (3) *Upper Aptian* (50 m: massive and thick bedded, jointed, stylolite, partly karistified limestone and dolomitic limestone).
– Middle Cretaceous epoch: It has two rock formation as follows: (1) *Albian* (200 m: well bedded marly limestone and shale grading with increased thickness of the limestone and marl upward); (2) *Cenomanian* (700 m: massive changing to thin bedded, highly fractured, jointed, and karistified dolomitic limestone and limestone with some thin beds of marly limestone and chert nodules).
– Upper Cretaceous epoch: This epoch with the two rock formations is composed mainly of silicate rocks, and it includes: (1) *Turonian* (200 m: moderately thick to thin bedded, marly limestone dominant chert nodules); (2) *Senonian* (400 m: marl and marly limestone, with massive, jointed, fractured and sometimes soft and friable rock facies).

3. Tertiary Period: Spreads over many localities in the coastal zone, Tertiary rocks are mainly situated in the North (near Chekka region), while they appear in two major parts, along the limbs of the Cretaceous anticline in the south.

– Eocene epoch: This is composed of (1) *Ypresian* (370 m: moderately thick bedded marl, chalky limestone and marly limestone); and (2) *Lutetian* (800 m: moderately thick to thin beds, highly fractured and jointed, and sometimes karistified with of Nummulitic limestone including chert nodules, interbedded with marly and chalky limestone).
– Miocene epoch: The Miocene has two major rock formations. These are: (1) *Burdigalian* (80 m: massive friable marly limestone and marl with silt and marl intervening); (2) *Vindobanian* (320 m: thick accumulations of conglomeratic limestone and sometimes clastic limestone).

- Pliocene epoch: It is a unique rock formation represented by the *Plaisancian* age (variable thickness, massive and reefal deposits of marly limestone and conglomerates with basaltic rocks).

4. Quaternary Period: The rocks of this period are distributed mainly in the coastal regions. However, the largest part of it is located in the Bekaa Plain.

- Pleistocene epoch: It is composed mainly of variety of alluvial deposits, soils and dunes.
- Holocene epoch: The majority of deposition is the marine, river and alluvial terraces.

2.4.2 Structures Geology

The geographic location of Lebanon along the eastern Mediterranean Basin, as well as its setting as an extension to the Red Sea Rift System, makes it a territory with tremendous tectonic activities. Thus, tectonism has a major role in the geological processes influenced the geologic setting of Lebanon, which can be described as complicated with predominant rock deformations.

Along the Tiberius Lake and the Dead Sea, the alignment of the Red Sea Rift System is sharply crossing the Lebanese territory where it typically observed near Yammounah area and then named as "Yammounah Fault".

According to Shaban and Hamzé (2017), the Lebanese territory including mainly Mount-Lebanon and Anti-Lebanon, comprises a regional folding system that cut by sets of nearly parallel strike-slip faults that resulting widespread fracture zones. Thus, Mount-Lebanon represents a regional monocline where folds and flexures of different dimensions occur among this structure. While, Anti-Lebanon represents elongated fold (i.e. anticline and syncline) and including several plunging anticlines.

1. Faults

The geological maps (scale 1:50000) produced by L. Dubertret (1953 and 1955) reveal several fault systems that span into different localities in Lebanon. However, there are two dominant fault types identified (from field survey and satellite images) in Lebanon. These are the gravity faults and wrench faults.

Thus, gravity faults (predominantly of the normal fault types) are widespread and they are mainly accompanied with vertical displacement and sometimes with a little diagonal motions. Hence, this type of faults is almost local with hundreds of meters movement (i.e. 400–500 m as measured in many outcrops) and result vertical blocking of different rock lithologies.

Wrench faults are also dominant; however, they cut for long distances (i.e. several kilometers). They are attributed to the local compression forces that are associated with small-scale strike-slip faulting, or with local jostling of blocks and squeezing of the sediments (Beydoun 1977). These are well pronounced in Mount-Lebanon ranges where equality in the lateral displacement of some lithologies along the wrench faults could be identified (Shaban 1987).

Fig. 2.6 Local anticline, a typical folding in Mount-Lebanon

2. Folds

As a results of consequent geological orogenesis over the past history, the Lebanese territory has been subjected to several compression forces that resulted a number of folding systems with different scales and mechanisms. Therefore, the major rock mass of the Lebanese territory represents a regional monoclinic structure of the Mount-Lebanon, while a large number of local folds are developed among this structure building up mountain chains (Fig. 2.6). Meanwhile, the entire Anti-Lebanon is almost a large-scale folding structure where many plunging anticlines and synclines exist.

The monocline of Mount-Lebanon is trending parallel to the Mediterranean Sea where its axial plan extends in the *NE-SW* direction. This monocline is restricted by the Yammounah fault to the east. Shaban (1987) stated that differential compression occurred as a result of regional rotational motion. The motion was accompanied with compressive stresses in the North and as a reverse reaction, tensile stresses in the South.

Flexures also occur in Mount-Lebanon where they often extend along to the inclined bedding plans. They are certainly developed in the coastal zone and the plateaus where these flexures are predominantly inclined seaward.

3. Fractures

The high rate of tectonism in Lebanon produces local deformations where fracture systems are originated. This can be resulted due to faulting and folding processes. For example, the presence of faults almost results fractures that develop in the proximity of these faults, as well as fractures are usually created on the hinge body of the anticlines. Hence, these deformations normally initiate karstification of the carbonate rocks.

Therefore, fracture systems in Lebanon are usually found as: (1) fissures with spacing of less than 1 m in general, and they often appear as irregular slickensides, dense spacing (i.e. usually one set with multiple plans); (2) joints where the displacement is not remarkable enough, and the average spacing is 10 cm and 2 cm for the hard and soft rocks; respectively.

2.5 Land Cover/Use

As a fundamental component of terrain characteristics and the related atmospheric conditions, land cover (LCU) must be considered. This is because LCU integrates between the natural and anthropogenic influencers where both act on many sectors, notably water and agriculture. Thus, LCU maps are often produced for many purposes, such as measuring agricultural areas, identifying the terrain changes, urban expansion, spread of green cover, etc.

The drawing of LCU maps are often elaborated from remote sensing applications where satellite images of different spatial and temporal resolution are used. Thus, terrain classification (e.g. supervised or non-supervised classification) is applied, and then it is followed by field truthening for the dedicated number of terrain components which must be primarily determined for mapping.

There are several methods and concepts applied to establish the LCU maps. However, the most used one is the CORINE Land Cover nomenclature, which is a vector map with a scale of 1:20 000. It has three levels hierarchical classification including levels 1, 2 and 3 for 5, 15 and 44 categories; respectively.

In Lebanon, several categorizations and methods have been done to produce the LCU maps for Lebanon over different time periods; and therefore, calculations were obtained on collective measures for terrain components (LCU) for Lebanon (examples, Masri et al. 2002; Darwich et al. 2004; SOER SOER 2010; SNC 2011; Hamzé and Faour 2014, Awad 2018).

The majority of the principal terrain features includes in a broad sense: bare lands, agricultural areas, forests, urban areas and water bodies. In Fig. 2.7, eight categories were established showing the main terrain components in Lebanon. Hence, the following component areas were calculated and their percentage to the entire Lebanon were also illustrated:

- Water bodies = 13.53 km^2 (0.13%)
- Urban settlements = 933.26 km^2 (8.92%)
- Brae soil and rock = 577.24 km^2 (5.52%)
- Fruit tress = 387.63 km^2 (3.70%)
- Forests = 1354.11 km^2 (12.95%)
- Shrubs = 1239.63 km^2 (11.86%)

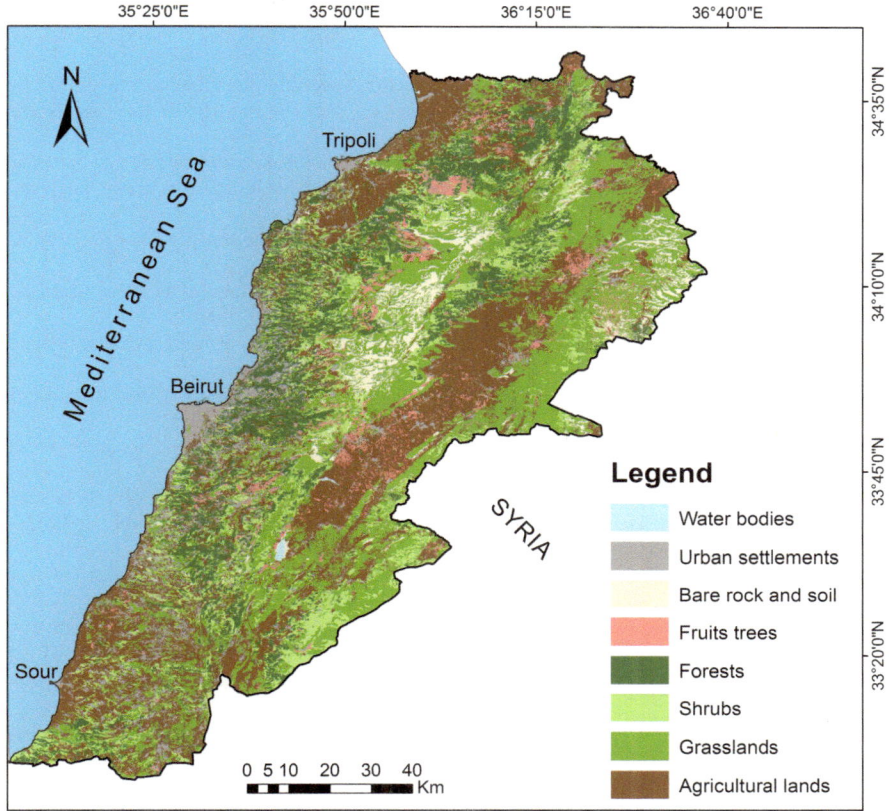

Fig. 2.7 Land cover/use of Lebanon (Awad 2018)

– Grass lands = 3191.31 km² (30.53%)
– Agricultural lands = 2755.21 km² (26.36%).

Chapter Highlights
– Rainfall rate ranges between 650 mm and 1500 mm, while it averages rainfall
  about 910 mm.
– Lebanon's climate can be described as sub-humid. In particular, Lebanon would
  not be remarkably affected by the drought.
– Lebanon has a unique topography makes it with different physiography from the
  Eastern Mediterranean regions.
– More than 60% of the carbonate rocks are karistified where several cavities, grot-
  tos and galleries have been identified.
– The oldest exposed rocks in Lebanon are attributed to the Middle Jurassic age
  where 17 major rock formations exist.
– More than 50% of the Lebanese territory is green, including agricultural lands,
  fruit trees, shrubs and forests.

References

Arkadan A (2008) Climate change in Lebanon: prediction uncertain precipitation events-Do climatic cycles exist? In: Climatic changes and water resources in the Middle East and North Africa, Springer, pp 59–71

Awad M (2018) Current and future trends in segmenting satellite images using hybrid and dynamic genetic algorithms. In: Bhattacharyya S (ed) Hybrid metaheuristics for image analysis. Springer, Cham

Awad M (2019) Toward precision in crop yield estimation using remote sensing and optimization techniques. Agriculture MDPI 9(54):1–13

Azar D (2000) Contribution à l'étude de la source karstique d'Afka. Mémoire de DEA, CREEN, Université Saint-Joseph, Liban, 104p

Beydoun Z (1972) A new evaluation of the petroleum prospects of Lebanon with special reference to the Pre-Jurassic. 18th Arab Pet. Cong., Algeria, 80(B-3)

Beydoun Z (1977) Petroleum prospects of Lebanon: re-evaluation. Am Assoc Petroleum Geol 61:43–64

Beydoun Z (1988) The Middle East: regional geology and petroleum resources. Scientific Press Ltd, London, 296p

Bou Kheir R, Girard M-C, Shaban A, Khawlie M, Faour G, Darwich T (2001) Apport de la télédétection pour la modélisation de l'érosion hydrique des sols dans la région côtière du Liban. Télédétection 2(2):91–102

CAL (1971) Atlas Climatique du Liban, Tome 1. Service Météorologique, Ministère des Travaux publics et Transports. Lebanon, Beirut

CAL (1973) Atlas Climatique du Liban, Tome 1. Service Météorologique, Ministère des Travaux publics et Transports. Lebanon, Beirut

CAL (1982) Atlas Climatique du Liban, Tome II. Ministère des publics et Transports, Service Météorologique, 40p

CAS (2015) Central Administration of Statistics. Bulletin, Environment and Agriculture

CHIRPS (2017) Climate Hazards Group InfraRed precipitation with station data. Available at http://chg.geog.ucsb.edu/data/chirps/

CNRS-L (National Council for Scientific Research, Lebanon) (2015) Regional coordination on improved water resources management and capacity building. Regional project. GEF, WB

Darwich T, Faour G, Khawlie M (2004) Assessing soil degradation by land use-cover change in coastal Lebanon. Lebanese Science Journal 5(1):45–60

DGAC (General Directorat of Civil Aviation) (1999) Rapport annuel. Beyrouth, Liban, 32p

Dubertret L (1933) La carte géologique au millionème de la Syrie et du Liban. Rev Géo Phys Géol Dyn 6(4):269–316

Dubertret L (1953) Carte géologique de la Syrie et du Liban au 1/50000me. 21 feuilles avec notices explicatrices. Ministère des Travaux Publics. L'imprimerie, Catholique, Beyrouth, 66p

Dubertret L (1955) Carte géologique de la Syrie et du Liban au 1/200000me. 21 feuilles avec notices explicatrices. Ministère des Travaux Publics. L'imprimerie, Catholique, Beyrouth, 74p

Edgell H (1997) Karst and hydrogeology of Lebanon. Carbonates Evaporites 12:220–235

Emberger E (1932) Sur une formule climatique et ses applications en botanique. La Météorologie:423–432

FAO (2009) AQUASTAT. Water report 34. Available at http://www.fao.org/nr/water/aquastat/countries_regions/LBN

Farhat N (2018) Effect of relative humidity on evaporation rates in Nabatieh region. Lebanese Sci J 19(1):59–66

Fawaz M (1992) Water resources in Lebanon, 1992

Geadah A (2002) Valuation of agricultural and irrigation water – the case of Lebanon. Water Demand Management Forumon Water Valuation, 25–27 June 2002. Lebanon, Beirut

Ghaddar N (1995) Climatological data. Monthly Bulletin. American University of Beirut, Lebanon

Guerre A (1969) Etude hydrogéologique préliminaire des karsts libanais. Hannon, Beyrouth 4:64–92

Hakim B (1985) Recherches hydrologiques et hydrochimiques sur quelques karsts méditer-ranéens: Liban, Syrie et Maroc. Publications de l'Université Libanaise. Section des études géographiques, tome II, 701p

Hamzé M, Faour G (2014) Space Atlas of Lebanon. National Council for Scientific Research, Lebanon, 121pp

Karam F, Breidy J, Stephan C, Rouphael J (2003) Evapotranspiration, yield and water use effi-ciency of drip irrigated corn in the Bekaa Valley of Lebanon. Agric Water Manag 63(2):125–137

Khawlie M, Shaban A (1998) Contribution of remote sensing studies to the fractured karstic coastal terrain, Lebanon-Water enhancement or hindrance. Conference on: flow, friction and fracture, AUB Center for Advanced Math. Sciences, Beirut, 1–7/7/1998

LARI (Lebanese Agricultural Research Institute) (2017) Climatic Data. Monthly Bulletin Department of Irrigation and Agro-meteorology (DIAM)

Lovallo M, Shaban A, Darwich T, Telesca L (2013) Investigating the time dynamics of monthly rainfall time series observed in northern Lebanon by means of the the detrended fluctuation analysis and the Fisher-Shannon method. Acta Geophys 61(6):1538–1555

Masri T, Khawlie M, Faour G (2002) Land cover change over the last 40 years in lebanon. Lebanese Sci J 3(2)

MOE/ECODIT (2002) State of the environment report, Lebanon. Prepared by ECODIT for the Ministry of Environment

MOEW (2010) National water sector strategy: baseline. MOEW, 15 September 2010

MoEW (Ministry of Energy and Water), UNPD (2016) National guideline for rainwater harvest-ing systems

N.C (2011) Lebanon's second national communication to the UNFCCC. Ministry of Environment. GEF. UNDP, 191pp

Na'ameh M (1995) Water problems of Lebanon. National Congress on Water Strategic Studies Center. Beirut (in Arabic), 67p

NOAA (National Oceanographic Data Center) (2013) Lebanon climatological data. Library. Available at http://docs.lib.noaa.gov/rescue/data_rescue_lebanon.html

Plassard J (1971) Carte pluviométrique du Liban au 1/200 000. Ministry of Public Works and Transport, Lebanon

Rey J (1954) Carte pluviométrique du Liban au 1/200000me. Ministère des Travaux Publics, République Libanaise

Sanlaville P (1977) Etude géomorphologique de la région littorale du Liban. Publications de l'Université Libanaise. Section des études géographiques, Beyrouth, tome I, 405 p

Shaban A (1987) Geology and hydrogeology of the Nabatieh area. Unpublished M.Sc. Thesis, American University of Beirut, Geology Department, 103p

Shaban A (2003) Etude de l'hydrogéologie au Liban Occidental: Utilisation de la télédétection. Ph.D. dissertation. Bordeaux 1 Université, 202p

Shaban A, Hamzé M (2017) Shared water resources of Lebanon. Nova, New York, p 150

Shaban A, Houhou R (2015) Drought or humidity oscillations? The case of coastal zone of Lebanon. J Hydrol 529(2015):1768–1775

Shaban A, Khawlie M (2006) Lineament analysis through remote sensing as a contribution to study karstic caves in occidental Lebanon. Revue Photo-interprétation AGPA Edition 4:2006

Shaban A, Abdallah C, Bou Kheir R, Jomaa I (2000) Conduit flow: an essential parameter in the hydrologic regime in Mount Liban. Karst 2000, Ankara, Turkey, 17–26/9/2000

Shaban A, Awad M, Ghandour A, Telesca L (2019) A 32-year aridity analysis: A tool for better understanding on water resources management in Lebanon. Acta Geophys 67(4)

SNC (2011) Second National Communication to the UNFCCC. Climate change vulnerability and adaptation. Ministry of Environment & GEF & UNDP, Beirut, Lebanon, 288pp

SOER (2010) The state and trends of the Lebanese environment. Ministry of Environment, UNDP, 355pp

Telesca L, Lovallo M, Shaban A, Darwich T, Amasha N (2013) Singular spectrum analysis and Fisher-Shanon analysis of spring flow time series: an application to Anjar Spring, Lebanon. Physica A 392(2013):3789–3797

Telesca L, Shaban A, Awad M (2018) Analysis of heterogeneity of aridity index periodicity over Lebanon. Acta Geophys 67(1):167–176

TRMM (2015) Tropical rainfall mapping mission. Rainfall archives. NASA. http://disc2.nascom. nasa.gov/Giovanni/tovas/TRMM_V6.3B42.2.shtml

UNDP (United Nations Development Program) (1970) Liban-Etude des Eaux Souterraines. United Nation, New York

WB (World Bank) (2003) Republic of Lebanon –Policy note on Irrigation sector sustainability. Rep. No. 28766 –LE

Chapter 3
Rivers

Abstract When Lebanon is described as the country with plenty water resources, this is usually inspired from the remarkable number of rivers spread on its territory. Even though, the Lebanese rivers are with small dimensions; however, it can be said that there is one river in each 750 km². The most creditable estimation of the discharge from these rivers is about 2800 million m³ per year, which constitutes a substantial part of the water balance in Lebanon. Less than 15% of this amount is exploited and the rest is either lost to sea or shared with the neighboring countries. Other than the domestic uses, rivers' water in Lebanon is used mainly for agriculture. Lebanon rivers are also used for hydro-power generation where is contributes to about 10% of electricity needs for the entire country. The discharge from the Lebanese rivers is sharply decreased and some rivers lost more than 60% of its average annual discharge. This can be attributed either to the direct abstraction from these rivers or pumping form groundwater reservoirs. In addition, the changing hydrologic regimes of the terrain surface plays a major role in controlling the amount of water in rivers. This chapter reveals a detailed discussion on Lebanese rivers including their watersheds, and even those for the major streams as well as the related geometric measures. This is also accompanied with quantitative estimations on the amounts of water in the Lebanese rivers.

Keywords Drainage system · Steams · Catchment area · Flow-meter · Dams · Shared water

3.1 Introduction

"Naher" in Arabic means river. This word is commonly mentioned in Lebanon for all water channels with permanent flow and even though it is applied for seasonal streams. Therefore, "Naher" is a widespread term for all linear water bodies which are densely distributed between different Lebanese regions.

However, not all linear water bodies can be considered as rivers if the hydrologic concepts are adopted. Thus, a "river" should be a water stream that discharges water

© The Editor(s) (if applicable) and The Author(s), under exclusive license to
Springer Nature Switzerland AG 2020
A. Shaban, *Water Resources of Lebanon*, World Water Resources 7,
https://doi.org/10.1007/978-3-030-48717-1_3

all year long. If this applied to the Lebanese rivers, few of them will be named as rivers. While, if we apply the concept that a "river" is a water channel where navigation can be made; therefore, Lebanese has only two or three rivers.

In general, the discharge of the Lebanese rivers is low enough if compared with international rivers. The average annual discharge from these rivers is about 3452 million m³. Hence, considering that only 25% of the discharge from Transboundary Rivers belongs to Lebanon, therefore, the net discharge of the Lebanese rivers will be about 2800 million m³/year.

If this estimation is divided on the 14 rivers of Lebanon that mean the average discharge will be 247 million m³ per year which is equivalent to the discharge of the Nile River in 1 day.

There is still a debate about the exact number of rivers in Lebanon. This is because most of these rivers, which were discharging water all year long in the past, became no longer perennial and dry over several months of the year.

The Lebanese rivers follow defined hydrologic regime which is to-tally controlled by its geomorphology. Therefore, Lebanon with its small area is considered as a regional hydrologic junction where water flows into three regional directions. These are regional flows: (1) northward to comprise a tributary of the Orates River Flow System, (2) southward forming a major tributary for Jordan River Flow System and (3) eastward where the Lebanese Coastal Rivers System flow to the Mediterranean Sea (Fig. 3.1).

3.1.1 Rivers Classification

Fourteen rivers is the number of rivers in Lebanon (Fig. 3.2). The adoption of this number depends on the morphometric characterization and catchment shapes of the existing perennial watercourses, even though this conflicts with their hydrology where water flow becomes intermittent. The Lebanese rivers have different orientations, and more certainly different flow directions and dimensions; and thus they can be classified as:

1. Coastal Rivers: These are the ten western rivers that are originated from Mount-Lebanon and trending from East to discharge in the Mediterranean Sea. They fed mainly from the snowmelt. The coastal rivers are short in length and the longest one is El-Awali River which is about 61 km (curved). These rivers almost have similar basin characteristics where the channel slope is relatively high and averages at 35–40 m/km, and this makes water flow with high rate which is estimated at 5–10 km/hour in average (Shaban and Hamzé 2017). This in turn shares in water loss into the sea.

2. Inner Rivers: These are four rivers which are totally characterized by different morphometry and hydrology including mainly flow direction and regime. One of these rivers (El-Kabir River) is originated from the most northern part of Lebanon and outlets into the Mediterranean Sea; two rivers (Al-Assi River and Litani River) are originated from the Bekaa Plain where the first flows northward to Syria and the second flows southward and then meandered towards the sea within

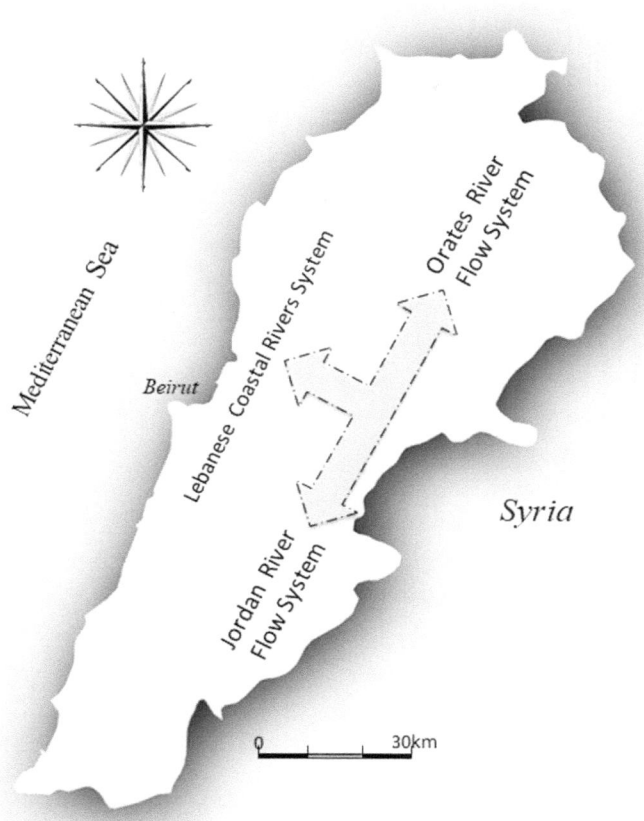

Fig. 3.1 Representation for the regional hydrologic junction in Lebanon

the Lebanese territory; while the fourth river (Hassbani-Wazzani River) is originated from Jabal Hermoun and then spans southward comprising the highest channel slope (40 m/km) of the four rivers. Thus, all these rivers are shared water resources except the Litani River.

3.1.2   Rivers Description

The 14 rivers of Lebanon are spread over small area of the country, and this makes it as a dense network of perennial watercourses (Fig. 3.2). Hence, these rivers are close to each other, and most rivers are at a distance of less than 10 km for each other.

The Lebanese rivers are mainly fed from springs/or snowmelt in where the later recharges groundwater reservoirs which in turn directly replenishes rivers or through

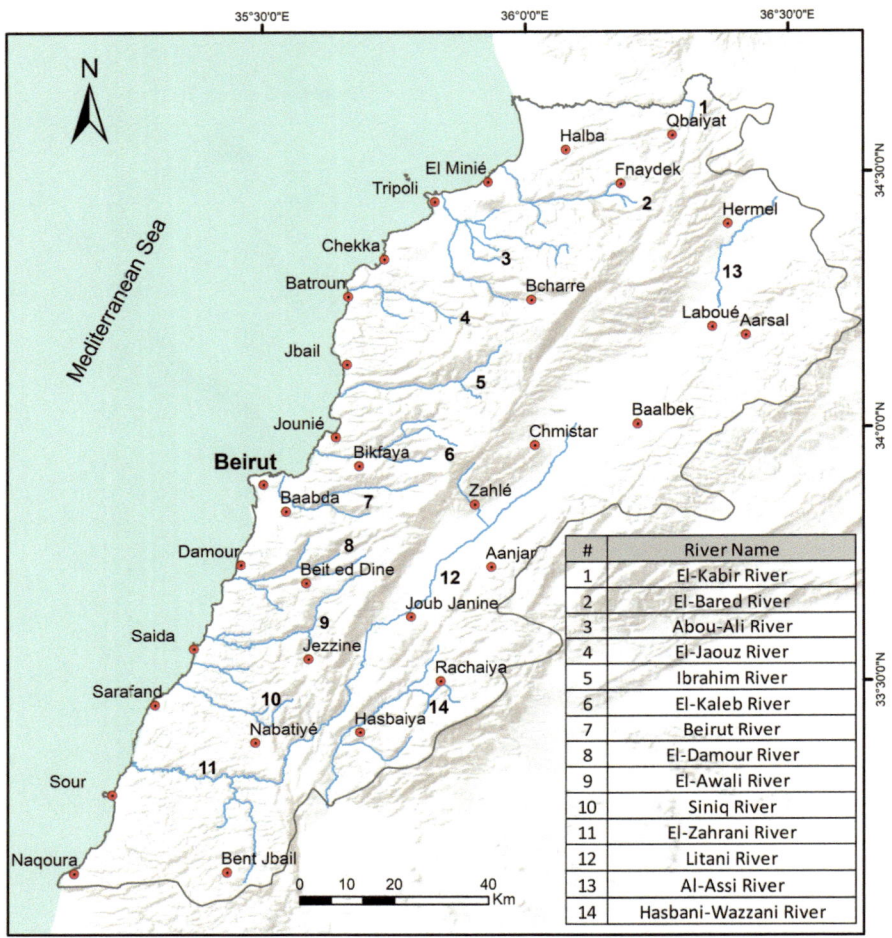

Fig. 3.2 Rivers of Lebanon

the located springs in rivers catchment. However, water supply from rainfall in rivers of Lebanon is almost minimal due to the steep slope of the catchment surfaces and the primary watercourses.

The following is a description for the 14 Lebanese rivers (almost from North to South). Detailed measures on these rivers will be discussed later on in this chapter:

1. El-Kabir River:
   This river flows water all year long, and it has 59 km length forming the boundary between Lebanon and Syria. It is commonly named as "El-Kabir El-Janoubi" which means the bigger-southern river since the major part of this river (i.e. >75%) in Syria. It is fed from more than 70 major springs where Naba'a Al-Safa Spring (North) is the largest one.

2. El-Bared River:

   It has a curved length of about 49 km and originated mainly from Ayoun As-Samak springs, as well as it is fed from three springs located in Jabal Al-Makmal where altitude of 2500 m is dominant.

3. Abou-Ali River:

   Even though this river has about 54 km length, yet it represents the largest coastal watersheds (468 km^2). It is also called Kadisha River because it is originated from Kadisha Grotto (Karstic spring) which flows water along Kanoubeen Valley and then Kadisha Valley.

4. El-Jaouz River:

   This river is fed from several springs in Tanoureen Al-Tahta and Tanoureen Al-Faouka area (> 2200 m). Hence, it has a curved length of about 37 km.

5. Ibrahim River:

   It is also named as Adounis River, and it discharges water over the entire year. It has the largest discharge rate among the coastal Lebanese rivers which averages about 495 million m^3 per year. This river extends about 50 km and it is fed mainly from Roueisat Spring and Afqa Spring which are located near Daher Al-Kadeeb Mountains, in Qartaba region.

6. El-Kaleb River:

   With about 41 km curved length, this river is fed from several springs (e.g. Ain Al Alak, Ain El-Karam, etc.) which are located near Jabal Sannine and Echoueir – Bekafya area.

7. Beirut River:

   It was named as Magoras River and it spans along 58 km where is originated from Naba'a Al-Ara'ar near Ba'abdat and from Naba'a Hammana and from many other springs near Jabal Al-Kneiseh and Jabal Sannine.

8. Ed-Damour River:

   Located directly south to Beirut, Ed-Damour River flows water over the whole year, and it has a length of about 54 km. It is originated from Naba'a As-Safa near Jabal Al-Akra, in Barook region (> 1200 m).

9. El-Awali River:

   It is the longest (curved) coastal river with 61 km. El-Awali River discharges water over the whole year, and it is originated mainly from Naba'a El-Barook Spring (1075 m) which is located in Jabal El-Barook and Jabal Niha. This spring supplies water for several villages in Esh-Shoof region.

   El-Awali River also receives water from Qanan Lake which is located near Jezzine, and this lake collects water that derived from the Qaraaoun Reservoir which is located in the Bekaa Plain.

10. Siniq River:

    This river is the smallest river in the coastal Lebanese rivers. Thus, it has the shortest length of about 21 km and 102 km^2 watershed. Also, it has the least discharge which averages about 60 million m^3/year.

11. El-Zahrani River:

    This perennial watercourse is fed mainly from Naba'a At-Tasseh Spring (910 m) which is located near Jabal Soujod. In addition, there are many other springs feed this river (e.g. Naba'a Kfarouh, etc.).

12. Litani River:

    This is the largest Lebanese rivers (entirely within Lebanon) where it has 174 km length, and it extends from the Bekaa Plain southward and then it divers seaward in the south. This river has also the largest catchment area of about 2110 km².

    There are several many springs (e.g. Shamsime, Anjar, Berdaouni, Koub Elias, etc.) feed the Litani River and thus water flows all year long. This river is significant water resources for Lebanon, and it supplies water to about one million people (Shaban and Hamzé 2018).

13. Al-Assi River:

    This is a transboundary water resources where it comprises a major tributary for the Orontes River. In Lebanon, Al-Assi River is about 33 km length within Lebanon and it flows water all year long where it is fed from several springs and mainly from Laboueh Spring and Ain Azarqa Spring.

14. Hasbani-Wazzani River:

    This is also a transboundary water resources where it represents a major water tributary for the Jordan River. It has a length of about 25 km, and it is fed from several springs located in Jabal Hermoun, and more certainly from Wazzani Spring.

3.2 Drainages Systems

Integrated water resources management (IWRM) requires characterization of a specific hydrologic system (e.g. watershed, aquifer, etc.). Typically, determining the dimensional measures and mapping of drainage systems is often considered as a perquisite step for further hydrological analysis and assessment of water flow regime. Hence, identifying the hydrology, morphometry and geometry is the primarily step done while proposing methods for watershed management including, for example, water harvesting, water supply, agricultural projects, hydro-power generation, etc.

3.2.1 Watersheds Extraction

The extraction of drainage system (streams network and their catchments area) is usually obtained from topographic maps, which are originally derived from aerial photographs. However, recent drainage systems can be extracted directly from stereoscopic satellite images where digital elevation models (DEMs) are generated by magnifying the pixel details, and then slopes can be extracted as a function of stream flow alignment for applying precise stream delineation. The accuracy of generated watersheds cartography depends on the scale of the used topographic maps or the precision and spatial resolution of the processed satellite images.

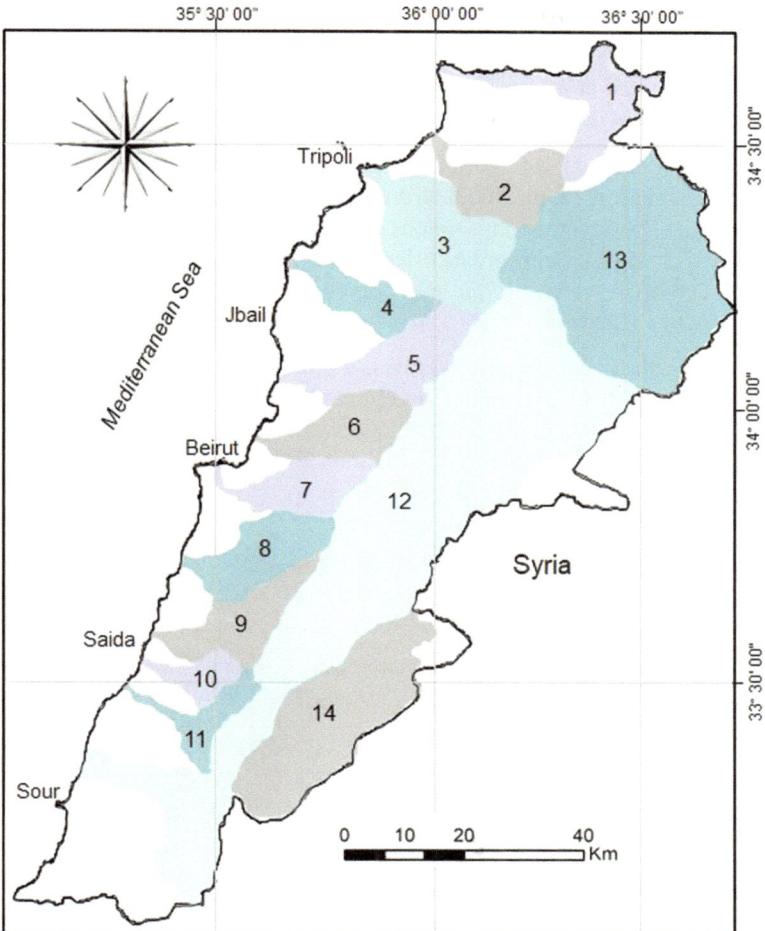

Fig. 3.3 Watershed of the Lebanese rivers

Shaban and Darwich (2010), put a detailed discussion about drainage systems of Lebanon after delineating all stream networks for the entire Lebanon at a scale of 1:200.000. This was elaborated by using topographic maps (scale 1:50.000) and satellite images (e.g. Aster and SRTM). In addition to the used tools, geographic information system (GIS) was also applied to help data manipulation, storage and maps production.

The tracing of drainage systems for Lebanon enabled determining water divides (i.e. watersheds) which were drawn through the connection of the highest points that surrounded the extracted drainage systems. Thus, the watershed map for Lebanon was produced (Fig. 3.3) and this was followed by numerous geometric, morphometric and hydrologic calculations.

3.2.2 Watershed Classification

Two principal parameters are considered to characterize surface water basins in Lebanon. These can be summarized as follows (Table 3.1):

1. Dimensional parameters: These are the dimensional aspect of water catchment including: shape, area, threshold altitude, estuary margin and slope gradient.
2. Hydrological parameters: This emphasis mainly on water flow permanency (i.e. perennial or intermittent stream), length of primary watercourse, number of branch outlets of water into the end-point (e.g. sea or lake, etc.) as well as the discharge rate (Shaban 2003).

Considering these parameters; however, watersheds in Lebanon can be divided into three types. These are: major, intermediate and minor watersheds. Hence, Lebanon occupies 21 major, 36 intermediate and 47 minor watersheds where 14 of these watershed belong to major rivers (Shaban 2003).

Table 3.1 Parameters of watershed classification in Lebanon

Parameter	Elements	Watershed type		
		Major	Intermediate	Minor
Dimensions	Prevailing shape	Funnel-like shape and L-shape	Elongated, funnel-like shape	Almost triangular
	Areal extent	Hundreds of kilometers	Tens of kilometers (<60 km^2)	Few tens of kilometers (<40 km^2)
	Altitude of threshold or headwater	>2500 m in North, 800–1000 m the South	1000–2500 in North, < 500 in South	Variable
	Estuary margin	1–5 km	1–4 km	1–10 km
	Slope gradient (*m/km*)	25–70 in North	10–50 in North	<10
		25–40 in South	10–30 in South	
Hydrology	Permanency of water flow	Mostly belong to perennial streams (4–8 months/year)	Seasonal streams (2–3 months/year)	Seasonal streams (several days)
	Length of major watercourse	>25 km	10–30	<10
	No. of sets outlet into the sea	1 set	1 set	Multi-sets
	Volume of discharged water	Tens to hundreds of millions of cubic meters	Several tens of millions of cubic meters	Usually less than five million m^3/ year

3.2.3 Geometric Measurements

Geometric measurements belong to the geometry of the outer boundary of a surface water basin (i.e. watershed), regardless of the characteristics of streams and tributaries among it. However, the extent of streams and tributaries as well as the existence of feeding water sources (i.e. springs), control the areal extent of a basins. Whereas major water basins (e.g. usually with considerable area) are often divided into subcatchments. This is dependent of the purpose of study or the applied assessment.

The areas of watersheds of rivers in Lebanon are shown in Table 3.2. It is clear that the average watershed area of rivers in Lebanon is 537 km^2, whereas it is about 252km^2 for the coastal rivers if excluding the Litani River. Hence, the largest one belongs to Litani River and the smallest one belongs to Siniq River.

1. Basin maximum length (L_m): This geometric variable is a function of the topographic orientation of a watershed where the maximum length of the primary watercourse is defined. It is related to the water flow velocity, and thus it controls the time of leakage, evaporation, and transpiration. It is calculated as a straight line along the main course from the highest point to the outlet. Table 3.2 shows the maximum length of watershed of the Lebanese rivers.
2. Basin width (W): Watershed length acts in opposite to its length and normally, watersheds with relatively large length need much more time to supply water from the upstream to the primary watercourse. Therefore, widths of the watershed of the Lebanese rivers are calculated, as shown in Table 3.2.

Table 3.2 Geometric measures of Watershed Rivers in Lebanon

No.	Watershed	A (Km^2)	L_m (Km)	W (Km)	E	F
1	El-Kabir River	195*	43	7.5	0.35	0.10
2	El-Bared River	277	27	11	0.70	0.38
3	Abou-Ali River	468	35	16	0.69	0.38
4	El-Jaouz River	196	32	6.5	0.49	0.20
5	Ibrahim River	326	40	8.5	0.57	0.20
6	El-Kaleb River	231	36	12	0.48	0.18
7	Beirut River	213	31	9	0.44	0.22
8	Ed-Damour River	333	32	10	0.65	036
9	El-Awali River	293	33	9.5	0.58	0.27
10	Siniq River	102	19	5.5	0.60	0.28
11	El-Zahrani River	142	28	6.5	0.48	0.18
12	Litani River	2110	145	16	0.36	0.10
13	Al-Assi River	1983*	51	31	0.98	0.76
14	Hasbani-Wazzani River	645*	52	11	0.55	0.24

*Watershed area within Lebanon

3. Elongation Index (E): It is the ratio between the diameter of the circle with the same area as the watershed, and the distance between the farthest two points in the watershed (Schumm 1956). It is expressed by the following equation:

$$E = \frac{2\sqrt{A}}{L_m \sqrt{\pi}}$$

Table 3.1 reveals the calculated E where Schumm (1956) classified the elongation ratio as: < 0.5, 0.5–0.7, 0.7–0.8, 0.8–0.9 and 0.9–1 for more elongated, elongated, less elongated, oval and circular; respectively.

4. Form Factor (F): It is the numerical index used to deduce ratio of the basin area to square of the basin length. It indicates the flow intensity within the watershed. Thus, form factors for watershed of the Lebanese rivers are shown in Table 3.2.

Hence, F should not exceed 0.7854. Therefore, smaller F value evidences more elongated, while high F value experience larger peak flows of shorter duration. According to Horton (1932) form factor is expressed as:

$$F = \frac{A}{L^2}$$

3.2.4   Morphometric Measurements

These measurements are referred to the dimensions, orientation and interconnection between streams in a watershed. They are function of streams origin and evolution, geomorphology and geology of the underlying stratum. Therefore, morphometry governs water flow regime and velocity, and thus it is usually accounted while applying assessments for surface water flow. Hence, the major morphometric measures of the Lebanese rivers' watersheds are (Table 3.3):

1. Relief gradient (R_r): This is the ratio of upland to lowland elevations within the catchment area, and it is expressed according to Pike and Wilson (1971) as:

$$R_r = \frac{\text{Mean Elevation} - \text{Minimum Elevation}}{\text{Maximum Elevation} - \text{Minimum elevation}}$$

2. Mean stream slope (S_c): It is calculated within the catchment area by dividing the difference in altitude between the source and the outlet over by the total stream length. Therefore, the higher the S_c the high flow rate along the main stream (i.e. run-off) and vice versa. It is represented by the following formula according to Raven et al. (2000):

$$S_c = \frac{\text{Elevation at source} - \text{Elevation at outlet point}}{\text{Length of stream}}$$

Table 3.3 Major morphometric measures of the Lebanese rivers' watersheds

No.	Watershed	Length (km) Straight	Length (km) Curved	R_r –	S_c m/km	S_b –	D S/km^2	M_r %	T_t –
1	El-Kabir River	46	59	0.34	17	25	3.00	78	3.77
2	Al-Bared River	37	49	0.25	13	14	1.00	76	2.75
3	Abou-Ali River	42	54	0.46	54	46	1.75	78	6.95
4	Ej-Jouz River	33	37	0.42	27	44	2.40	89	5.51
5	Ibrahim River	44	50	0.47	63	45	5.30	88	16.30
6	El-Kaleb River	35	41	0.57	66	45	6.10	85	17.25
7	Beirut River	48	58	0.53	52	50	5.80	83	14.89
8	Ed-Damour River	45	54	0.51	51	46	5.30	83	18.15
9	El-Awali River	50	61	0.33	36	44	4.35	82	10.90
10	Siniq River	18	21	0.26	7	9	2.35	86	7.54
11	Ez-Zahrani River	36	41	0.28	8	13	2.10	88	8.16
12	Litani River	163	174	0.21	5	8	0.84	83	0.54
13	Al-Assi River	31	33	0.25	19	21	1.27	94	13.03
14	Hasbani-Wazzani River	22	25	0.27	14	11	1.14	88	5.06

3. Mean catchment slope (S_b): This represents the difference in elevation between defined points in the catchment over length of the catchment. The formula of Sb was plotted by Morisawa (1976) as follows:

$$S_b = \frac{\left(\text{Elevation at } 0.85\,L\right) - \left(\text{Elevation at } 0.10\,L\right)}{\text{Elevation at } 0.75\,L}$$

Where L is the maximum length of the watershed, and measurements are taken along this line (0.10 L near the lower part of the catchment, 0.85 L towards the upper end). Whereas, the slope (in degree) = $\tan^{-1}$ (slope in decimal form). Hence, the increased Sb is a function of high rate of the overland flow and vice versa.

4. Drainage density (D): This is expresses the sum of streams within a specific area within the watershed, and thus higher stream density reflects low permeability of terrain surface and vice versa (Shaban et al., 2004). Therefore, it can be calculated according to the following equation:

$$D = \Sigma \frac{L\left(\text{total of all stream segments}\right)}{A\left(\text{area of the basin}\right)}$$

5. Meandering ratio (M_r): It express the ratio straight and curved length of the primary stream in the watershed (Shaban and Hamzé 2018). Hence, higher M_r reflects low run-off and exceeded sedimentation. M_r can be calculated as follows:

$$M_r = \frac{L\left(\text{straight}\right)}{L\left(\text{curved}\right)}$$

6. Texture topography (T_t): The texture topography indicates the tendency of a terrain to percolate water as a result of the lithology and rock structures of the basin terrain. It defines the total number of streams (N_s) of all order in a basin per perimeter (B_p) of the basin (Horton, 1945). Hence, texture topography is calculated by the following equation (Smith, 1950):

$$T_t = \sum N_s / B_p$$

According to Smith (1950), texture topography can be classified as: very coarse (<2), coarse (2–4), moderate (4–6), fine (6–8) and very fine (>8).

Table 3.3 Reveals the texture topography of the Lebanese rivers watersheds after considering the topographic maps of 1:50.000 scale for calculating the streams number.

3.3 Quantitative Measures

In order to investigate the hydrologic properties that characterize the surface water-capturing systems (i.e. watersheds), water input and output in the system must be primarily calculated. This can be done following several approaches of quantitative analysis. Normally, data from gauging stations are utilized to measure the quantitative elements of water that enters into, and outlets from, the watersheds. This must be accompanied with calculating dimensional and geomorphological measures using mainly topographic and geologic maps, in conjunction with satellite images processing and GIS applications.

As an integral figure of the hydrologic cycle, the input/output of the surface water system (watershed) can be calculated by measuring the volume of precipitated water enters the system and the volume of water outlets from this system (i.e. river mouth).

Data to calculate the water volume that enters the watersheds were taken from the available time series from meteorological ground stations. Also, there was essential contribution from remotely sensed products. Thus, data was retrieved from Tropical Rainfall Mapping Mission – TRMM (2015); Climate Hazards group Infrared Precipitation with Stations – CHIRPS (2017); and from National Oceanographic Data Center – NOAA (2013).

For the discharge (output), the majority of stream flow data for the Lebanese rivers was adopted from database delivered by the Litani River Authority (LRA 2017). This data is available on daily basis with a unit of m^3/sec, as well as in million m^3 per year.

In order to attain an Effective Uniform Depth (EUD) of rainfall for the watersheds of the Lebanese rivers, the "*Isohyet*" method was applied. Therefore, water volume enters each watershed was calculated (Table 3.4).

The results are used for further comparative analysis for each watershed. Such analysis is helpful for water resources assessment and management, including:

Table 3.4 Water volumes enter and outlet from the Lebanese rivers' watersheds

No.	Watershed	Area (A) Km²	Rainfall (R) Mm³/year	Discharge (D)	D/R %	R/A Mm³/km²
1	El-Kabir River[a]	303	260	222	–	0.38
2	Al-Bared River	284	225	165	73	0.79
3	Abou-Ali River	482	505	365	72	1.04
4	Ej-Jouz River	196	125	80	64	0.64
5	Ibrahim River	326	380	495	131	1.16
6	El-Kaleb River	237	330	225	66	1.39
7	Beirut River	216	260	100	38	1.20
8	Ed-Damour River	333	335	255	76	1.00
9	El-Awali River	291	320	280	88	1.09
10	Siniq River	102	100	60	60	0.98
11	Ez-Zahrani River	140	145	200	137	1.03
12	Litani River	2110	2078	360	17	0.98
13	Al-Assi River[a]	1930	1254	420	–	0.65
14	Hasbani-Wazzani River[a]	645	598	225	–	0.89

[a]Shared watersheds and the illustrated data for the area of the watersheds in Lebanon

water losses, consumption, supply/demand ration and monitoring system for climate changes.

In this study, water volume was measured for all major watershed types, the ones were considered, as they encompass large areal extents. Table 3.4. This table shows the water volume as an input and output for the watershed systems. Hence, it is obvious that the average volume water inputs is 480 million m³/year, and the average outputs is 247 million m³/year. Therefore, the Lebanese rivers, in general, discharge about 51% of the precipitated which is generally large enough considering that the rest accounts for evapotranspiration and groundwater recharge into substratum. This is attributed to the following:

1. The steep slope, which accelerates the water flow energy, thus reduces the time for water recharge,
2. The existed rainfall patterns which has been lately charged to torrential rain, which is in turn make the run-off very rapid,
3. The terrain surfaces of most watershed where the infiltration rates have been reduced due to exceeded urban settlements,
4. The existence of snow, which is covers about 25% of Lebanon. Hence, snowmelt contributes uniformly in feeding surface and subsurface water sources.

The relationship between discharge and rainfall (D/R), as well as between rainfall and area (R/A) were calculated and illustrated in Table 3.4 and Fig. 3.4. However, (D/R) for shared watersheds were not calculated, because the areas of these watersheds extend outside Lebanon where many other hydrologic element play a role.

Table 3.4 shows that among the coastal watersheds, some D/R values are anomalous. This is well pronounced in watersheds of Ibrahim River and Ez-Zahrani River,

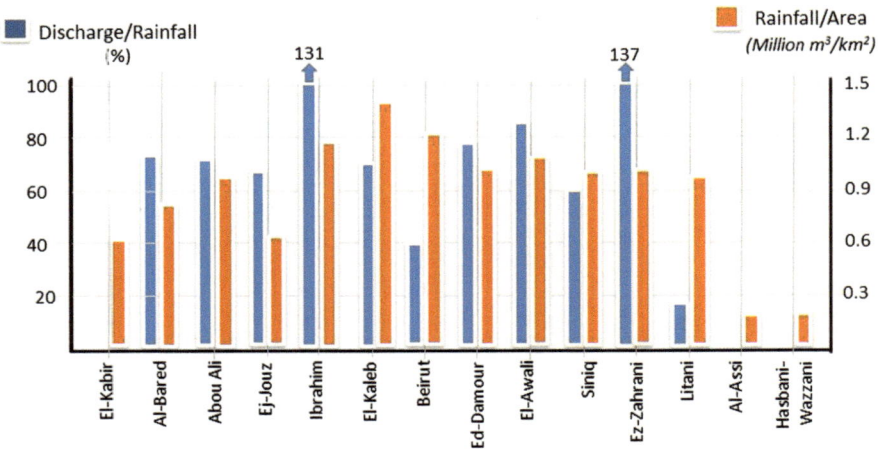

Fig. 3.4 Comparison between area, rainfall and discharge form Lebanese rivers

where the discharge rate exceeds the volume of precipitated water and this can be attributed mainly to the groundwater feeding from the aquifers spreading beneath different watersheds (Shaban et al., 2009). Besides, some other major watersheds discharge less water volume with respect to the precipitated water, such as Beirut River and Litani River. However, the D/R values, in general, indicates higher run-off in the watersheds of the coastal rivers in Lebanon (Fig. 3.4).

For the ratio of the rainfall with respect the area, it is obvious that he average ratio is 0.94 million m^3 per km^2. Which means that the average rainfall rate is about 940 mm is the average rainfall rate for all major watersheds (i.e. rivers' watersheds) in Lebanon.

The correlation between rainfall peaks and the amount of water discharging into the sea from the Lebanese coastal rivers was established using remotely sensed data applied by Shaban et al., (2009). For this correlation, MODIS-Terra satellite images and TRMM radar data were used. Therefore, daily rainfall measures from TRMM were compared with the areal extent of water plums occur into the sea as a results of rainfall/discharge regime.

The areal extent of water plumes into the sea was converted into water volume. This concept behind this conversion was dependant on considering the water plumes for rivers with known water discharge into the sea during regular stream flow periods. Hence, the applied method, to calculate the discharging water volume into the sea, can support measures taken from ground instruments (i.e. flow-meters). The applied method is also characterized by low cost and it reduces time consuming.

The steep sloping terrain (80 m/km in average) as well as the short length of rivers (almost <60 km) result relatively rapid stream flow towards the sea. According to Shaban et al., (2009), the average period needed for water to flow from top mountains until it reaches the sea is often less than 5 hours. While, the average lag time (i.e. time between rainfall peak and appearance of stream water into the sea as plumes) was estimated at 2.4 days for all coastal rivers.

The stream flow from the coastal rivers is often tempered by dense rock fractures and sub-surface karstification, and moreover by the higher meandering of valley courses.

3.4 Transboundary Rivers

There are three rivers in Lebanon that shared with the neighboring countries (i.e. Transboundary Rivers). These are: El-Kabir River, Al-Assi River and Hasbani-Wazzani River. These rivers are often given concern, because the largest portion of water in these rivers is flows outside Lebanon without efficient utilization.

According to Shaban and Hamzé (2017), the quota of Lebanon from water in these rivers is not satisfactory, and thus water benefit can be described as poor to moderate, moderate and poor for El-Kabir River, Al-Assi River and Hasbani-Wazzani River; respectively.

There transboundary rivers encompass permanent stream flow all year long, and thus, the total stream flow of the rivers is about 867 million m^3 per year which is equivalent to 25% of the total discharge from the Lebanese rivers.

3.4.1 El-Kabir Transboundary River

This is the upper northern coastal river in Lebanon, where it represents the political border (59 km length) between Lebanon and Syria. The river has a watershed area of about 972 km^2, where Lebanon occupies about 303 km^2 (31%) of it (Fig. 3.5). The average annual rainfall rate ranges between 800 and 900 mm year, thus provides water within the river catchment of about 260 million m^3/year.

A number of springs, which are fed mainly from snowmelt, provide water to El-Kabir River. These are: Naba'a Es-Saffa Spring and other karstic and fault springs within Lebanon, and Ain Essaqa, Ain Echrchara and Ain Samawae Springs which are located in Syria (Shaban and Hamzé, 2017).

The estimated percentage of water use, from El-Kabir River, ranges between 10 and 12% (<27 million m^3/year) from the Lebanese part, while the rest is either used from Syria or flows into the sea.

3.4.2 Al-Assi Transboundary River

"Al-Assi River", as domestically called in Lebanon, represents the southern tributary of Orontes Rivers with 33 km length (Fig. 3.5). The later shared Lebanon. Syria and Turkey where it outlets in the Mediterranean Sea.

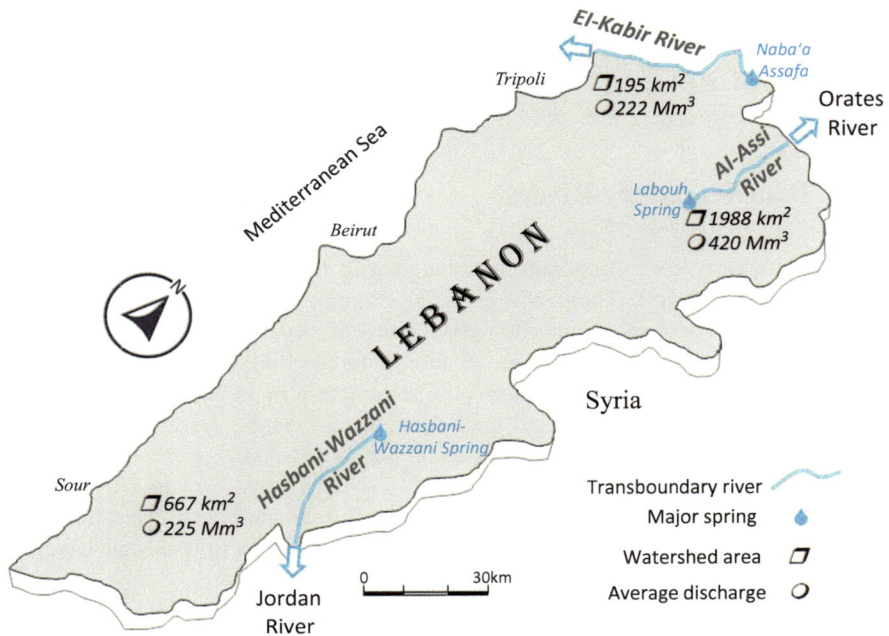

Fig. 3.5 Transboundary Rivers of Lebanon

Orontes River encompasses a regional catchment which is about 25,300 km². Hence, Lebanon occupies only about 1930 km² (7.5%). Even though the average annual discharge from this river is 736 million m³, yet the discharge of Al-Assi River is about 420 million m³/year.

The average rainfall rate over Al-Assi River (as a part of Orontes River) is about 650 mm. This contributes essentially to the stream flow of the river. In addition, there is a number of major karstic and fault springs that feed the river, and more certainly the Labouh Spring. In addition to springs of Ain Zarka Ain Flikah, Ras Baalbek, Eshawagheer and Azeraa springs (Shaban and Hamzé, 2017).

For Al-Assi River, the estimated percentage of water used from the river does not exceed 15%, while the remaining amount continuous flowing towards the north to join Orontes River.

3.4.3   Hasbani-Wazzani Transboundary River

Located in the south eastern side of Lebanon and flows towards the Palestinian Territories and Jordan, the Hasbani-Wazzani River has a 25 km where it originated from Jabal Hermoun (in Lebanon). It comprises the northern tributary of Jordan

River, which owns a catchment area of about 18,425 km^2; while it occupies only 645 km^2 in Lebanon.

Hasbani-Wazzani River is fed from several karstic springs where Wazzani Spring is the principal one, and it is integrally receiving water from the melting snow. While the average rainfall rate over the river catchment is about 980 mm (Shaban and Hamzé, 2017).

Chapter Highlights

- Lebanon occupies 21 major, 36 intermediate and 47 minor watersheds where 14 of these watershed belong to major rivers.
- The net average annual discharge from the 14 Lebanese rivers is about 2800 million m^3.
- Geometric and morphometric measurements have been calculated for the 14 catchments of the Lebanese rivers.
- The average period needed for water to flow from top mountains to the sea is often less than 5 h.
- The average lag time, between rainfall peak and appearance of stream water into the sea as plumes, was estimated at 2.4 days for the coastal rivers.
- The average annual discharge from the three transboundary rivers is 867 m^3.

References

CHIRPS (2017) Climate Hazards Group Infrared Precipitation with Station data. Available at: http://chg.geog.ucsb.edu/data/chirps/

Horton RE (1932) Drainage-basin characteristics. Trans Am Geophys Union 13:350–361

Horton R (1945) Erosional developments of streams and their drainage basins: hydro-physical approach to quantitative morphology. Geol Soc Am Bull 56:275–370

LRA (Litani River Authority) (2017) Rivers discharge records database (Unpublished Report)

Morisawa M (1976) Geomorphology laboratory manual. Wiley, New York, pp 1–253

NOAA (National Oceanographic Data Center) (2013) Lebanon climatological data. Library. Available at: http://docs.lib.noaa.gov/rescue/data_rescue_lebanon.html

Pike R, Wilson S (1971) Elevation-relief ratio. Hypsometric integral and geomorphic area-altitude analysis. GSA Bull 82:1079–1084

Raven, P., Holmes, N., Naura, M., Dawson, F. 2000. Using river habitat survey for environmental assessment and catchment planning in the U.K. Hydrobioloigia Vol. 422(0):359–367

Schumm S (1956) The elevation of drainage systems and slopes in badlands at Perth Amboy, New Jersey. Geol Soc Am Bull 67:597–646

Shaban A (2003) Studying the hydrogeology of occidental Lebanon: utilization of remote sensing. Ph.D. dissertation. Bordeaux 1 Université, 202p

Shaban A, Darwich T (2010) Mapping watersheds for sustainable management in Lebanon research project. CNRS-L, Final report, 23 pp

Shaban A, Hamzé M (2017) Shared water resources of Lebanon. Nova Publishing, New York, p 150

Shaban A, Hamzé M (2018) The Litani River, Lebanon: an assessment and current challenges. Springer, 179 p. https://doi.org/10.1007/978-3-319-76300-2

Shaban A, Bou Kheir R, Khawlie M, Froidefond J, Girard M-C (2004) Caractérisation des facteurs morphométriques des réseaux hydrographiques correspondant aux capacités d'infiltrations des roches au Liban occidental. Zeitschrift fur Geimorphologie 48(1):79–94

Shaban A, Robinson C, El-Baz F (2009) Using MODIS images and TRMM Data to correlate rainfall peaks and water discharges from the Lebanese Coastal Rivers. J Water Resour Protect 4:227–236

Smith K (1950) Standards for grading texture of erosional topography. Am J Sci 248:655–668

TRMM (2015) Tropical rainfall mapping mission. Rainfall archives. NASA. http://disc2.nascom.nasa.gov/Giovanni/tovas/TRMM_V6.3B42.2.shtml

Chapter 4
Springs

Abstract A major aspect of water resources, springs are widespread in Lebanon, and they are characterized by different hydrogeological mechanism, and thus by diverse flow rate and regime. The natural setting of Lebanon with its complicated geology, and more certainly the dominant rock deformations that interrupt the lithostratigarphic sequence, gives a chance for groundwater to seep on terrain surfaces as surface water flow. As per their hydrogeological linkage, it is not precise to distinguish water in springs from those in rivers and groundwater, yet these springs occupy an essential part of water budget in Lebanon and they discharge considerable water volume estimated at about 1410 million m³/year, which is equivalent to about 36% of rivers' water. There are about 1800–2000 major springs in Lebanon, which are attributed mainly to karstic and fault springs type. The largest part of them is considered as the primary source of water for rivers, and all rivers in Lebanon are substantially fed from springs where, in many instances, one or two springs fulfill the stream flow all year long in these rivers. The majority of water in these springs is derived from snowmelt that accumulated on the mountainous regions of Lebanon (Shaban 2003). This chapter will give a detailed discussion on the springs of Lebanon including their lithostratigraphy and rock structures controls as well as the discharge regime for springs locate on terrestrial environment an even those off-shore.

Keywords Snowmelt · Steam flow · Percolation · Impermeable rocks · Seeps · Fault spring · Karstic conduits

4.1 Introduction

It is not a surprisingly that most of the major springs are found in/or nearby urban clusters. This is because over the past history, springs were initially the natural attractive element where people select their settlements and then utilize their needs of water. This gives a special feature for Lebanon where several springs are discharging, and they are known as the main sources for potable water, heritage and

A. Shaban, *Water Resources of Lebanon*, World Water Resources 7,
https://doi.org/10.1007/978-3-030-48717-1_4

even touristic localities. Therefore, considerable amounts of water from springs are used to balance and secure water demand.

According to METAP (1995), there are about 2000 springs in Lebanon with an average discharge rate of about 12 l/s, and thus the total water volume from springs is estimated at 630 million m^3. Besides, MoEW (2010) stated that the total yearly yield from springs exceeds 1200 million m^3. Hence, less than 200 million m^3 is available during the summer period.

MoEW and UNDP (2014) reported that 5050 springs are spread in Lebanon. These springs were identified from topographic maps (1:20.000 scale). There are 2290 of these springs are named either as "Ain" or "Naba'a". The biggest number of these springs is located in clusters (i.e. considerable number of springs in small locality) where the geology governs their geographic distribution. Therefore, the majority of these clusters is found to be in the upper and middle parts of Mount-Lebanon where intensive rock deformations and karstification of carbonate rock are dominant.

Some of these springs in Lebanon are characterized by permanent discharge all year long, whereas abrupt seasonal flow variations occur. Besides, other springs do not yield water in dry seasons and other ones has recently dried.

The dryness of many springs has been exacerbated lately in Lebanon as a result of the increased population and the oscillating climatic conditions. Therefore, most of the Lebanese springs lost significant portion of their water and some of them are showing decline in the discharge rate reached to more than 40% over the last five decades (Shaban 2011).

The hydrogeology of springs in Lebanon implies their occurrence in the carbonate rocks (i.e. limestone, dolomitic limestone and dolomite), which are characterized by intensive fracturing systems and the development of karstification. The concept behind groundwater seeps on terrain surfaces (i.e. springs) includes the percolation of water into the carbonate rocks which are usually interbedded with impermeable stratum (e.g. marl and clay), and then resulting an overland flows of groundwater along the contacts between the bedding planes. In addition, the dissolution of the carbonate rocks (karstification) gives a chance for groundwater to flow on terrain surface which are governed by the characteristics of reservoir rocks (Example Fig. 4.1).

Nevertheless, springs in Lebanon are also found in the clastic rock units (e.g. sandstone), and some others are discharging from the loose materials of rocks and soil (i.e. alluvial and colluvial deposits).

In addition, there is a unique hydrgeologic phenomenon occurs as groundwater discharge into the marine environment of Lebanon. It is the existence of sub-marine springs which are widespread along the shoreline of Lebanon where karstic conduits and faults transmit groundwater into the sea.

Large number of springs in Lebanon are used for irrigation while other major ones used for domestic water supply. Thus, most of these springs are under the mandatory of MoEW where LRA and the four regional Water Establishments (WE) manage these resources. Meanwhile, private springs (i.e. located in private lands) are few enough or sometimes with minimal discharge.

Fig. 4.1 Khrayzat Spring, a typical spring issuing from karstic conduits

LRA is the principal entity who is in-charge to maintain and manage gauging stations (i.e. flow-meters) for springs (as well as rivers) in Lebanon. Therefore, long time series of flow measurements for the major spring are available since 1960s. However, continuous records are rarely available. In addition, most flow-meters fixed on springs do not provide representative measures of the discharge, given that they often fixed downstream of the spring, and do not take into account the upstream water that is being directly pumped by the inhabitants (MoEW and UNDP 2014).

4.2 Springs Characteristics

Based on the concept that springs are fed from groundwater, these resources are always considered as an aspect of groundwater. However, the author has several studies where he included springs as surface water resources (Shaban 2003, 2009a, b, 2011, 2014). This is because they are visible resources, and thus they can be easily measured, managed and tapped.

Springs are usually characterized either according to their hydrogeology and the mechanism of water flow and then it is called "spring type", or due to their discharge and flow continuity where they are classified according to their discharge rate.

In Lebanon, there is an obvious concern given to springs because they significantly contribute for water supply. Thus, many studies were done to investigate springs (Guerre 1969; Heybrook 1969; Abbud and Aker 1986, Edgell 1997; Azar 2000, etc.).

4.2.1 Types of Springs

Water flow from springs are driven by hydraulic heads that are governed by the existing hydrological forces. Thus, the hydrogeological setting from where springs discharge water are different. In this respect, there are many factors controlling the hydrologic behavior of springs including mainly the precipitation rate, volume and water level of groundwater and existence of surfaces of weakness within the reservoir rocks.

There are several classifications for springs including mainly topography, aquifer type, discharge, flow direction, chemical characteristics and temperature. Yet, mechanism of flow is still the principal factor, which is controlled by the existing geologic of the spring locality. Thus, five major types of springs occur in Lebanon. These are:

1. Contact springs: This type of springs is also called "ordinary springs", and they occur along the interfaces between different geological formations where porous and permeable rocks are overlain impermeable and compacted rocks.

   These springs are dominant in Lebanon (Table 4.1), where the fractured and karistified limestone and dolomite rock masses are commonly interbedded with impermeable marly limestone, marl and argillaceous rocks. Thus, the resulting contact gives a chance for water to emerge along the interface between both.

   Contact springs have several aspects where the interbedding and the inclination of rock layers have a major role, and this controls the flow mechanism from these springs. For example, contact springs are described as "overflow" springs when the level of water table exceeds the level of the contact between two rock layers with different lithological characteristics.

   Figure 4.2 shows a hydrgeologic section of a contact springs (overflow aspect) in Lebanon where Ain Ez-Zarka near Hermel is a typical example. It occurs at the bottom of massive limestone and dolomitic limestone of the Cenomanian-Turonian rocks which are tilted towards the Bekaa Plain, and then facing the Neogene rocks which are dominant with marly limestone. Therefore, the estimated potential yield of this spring is about 78 million m^3/year.

2. Karstic springs: These springs are probably characterized by the largest discharge of groundwater from a single opening, which is usually a cave-like shape. Thus, the run of groundwater follows sub-surface conduits and galleries where it spans, in many instances, for several kilometers. Hence, the flow of karstic springs (i.e. pipe flow) reveals abrupt changing in the opposite of other springs type where groundwater seeps as diffusion flow among porous and permeable rocks.

Table 4.1 Major springs of Lebanon

No	Spring	Discharge (m^3/s)[a]	M[b]	Spring type	Related rock formation[c]
1	Afqa	4.62	1st order	Fault/Karstic	C_4
2	Jeitta	4.48		Karstic	J_6
3	Rouaiss	3.55		Fault	C_4
4	Adoniss	3.37		Fault	C_4
5	El-Aarbaain	3.11		Fault	C_4
6	Yamouneh	2.84		Fault/Karstic	C_4
7	Ain Ez-Zarqa	2.44	2nd order	Contact	C_4
8	Ain Ed-Deleb	2.20		Karstic	C_1
9	Anjar	2.00		Contact	C_4
10	Dalle	1.93		Fault	J_6
11	Wazzani	1.86		Contact	J_4
12	Aqoura	1.70		Karstic/Fault	C_4
13	El-Barouk	1.58		Contact	J_4
14	Al-Assal	1.43		Fault	C_4
15	Al-Safa	1.42		Contact	J_4
16	Berdaouni	1.41		Karstic	C_4
17	Habb	1.24		Artesian	C_4
18	Hasbani	1.21		Karstic	J_4
19	Shoukkar	1.15		Contact	C_4
20	Rachaaine	1.10		Contact	C_4
21	Sreid	1.01		Fault	C_4
22	Labouh	0.95		Karstic	C_4
23	Ras Al-Ain (Sour)	0.88		Artesian	E_2
24	Al-Safa (North)	0.87		Karstic	C_4
25	Edardara	0.80		Karstic	C_4
26	Yanabiaa El-Hermel	0.78		Karstic	C_4
27	Mar Sarkiss	0.76		Karstic	C_4
28	Kashkoush	0.75		Karstic	J_6
29	Al-Laban	0.74		Fault/Karstic[d]	C_4
30	Kadisha	0.72		Karstic	C_4
31	Chaghour	0.72		Karstic	C_4
32	Aamiq	0.71		Karstic	J_6
33	Bkeftine	0.70		Karstic	C_4
34	Qeb Elias	0.69		Karstic	J_6
35	Machghara	0.67		Karstic	E_2
36	El-Hadid	0.67		Karstic	C_4
37	El-Qbayat	0.66		Karstic	C_4
38	Yanabih	0.64		Karstic	C_4
39	Ain Zehalta	0.61		Karstic	J_6
40	Sannine	0.58		Contact	C_4
41	Fawar Entelias	0.56		Artesian	C_4
42	Khrayzat	0.54		Karstic	J_6

(continued)

Table 4.1 (continued)

No	Spring	Discharge (m^3/s)[a]	M[b]	Spring type	Related rock formation[c]
43	Nahle	0.52		Karstic	C_4
44	Moghr Toffaha	0.50		Karstic	C_4
45	Et-Tasse	0.49		Karstic	C_4
46	Jezzine	0.48		Contact	J_6
47	Chamsine	0.46		Contact	C_4
48	Chtcura	0.46		Fault	J_6
49	Ahrous	0.45		Karstic	C_4
50	El-Kadi	0.41		Artesian	C_4
51	El-Jouz	0.38		Fault/Karstic	C_4
52	Al-Abain	0.37		Contact	C_4
53	Ras Al-Ain (Baalbak)	0.35		Artesian	C_4
54	Mar Sema'an	0.34		Fault/Karstic	C_4
55	Cheba'a	0.32		Fault	J_6
56	Na'ass	0.30		Contact	J_6
57	Ain El-Baida	0.26	3rd order	Fault/Karstic	C_4
58	Aayoun Es-Samak	0.23		Artesian	E2
59	Yahfoofa	0.23		Contact	C_4
60	Ras Al-Ain (Terbol)	0.22		Fault	m_1-m_2

[a]Average discharge in wet and dry seasons
[b]M: Meinzer Classification (1923)
[c]Major rock formation feeding the spring (According to Table 2.3 in Chap. 2)
[d]Almost fault spring with karstification

It is well known that karstic springs have no defined networking since they follow the sub-surface dissolution routes (e.g. conduits, shafts, galleries, swallet, etc.); thus groundwater often flows chaotically through these underground openings and their flow regime is difficult to be delineated.

This type of springs is the most common one in Lebanon (Table 4.1), and it can be estimated that between 50% and 60% of the Lebanese springs are of the karstic type, notably that the largest part of the geologic section of Lebanon is composed of carbonate rocks.

Karst springs in Lebanon are found, in many cases, combined with other rock deformation systems. This is normal since the existing rock deformation are the initial stage for karstification. A good example can be represented by Jeitta grottos spring (Fig. 4.3), which is a typical karstic spring (4.48 m^3/s) in Lebanon, and it is associated with folding and faulting systems (Shaban 2003).

3. Fault springs: This aspect of rock deformation results elongated surface of weakness as well as fragile zones (i.e. stratum with high fissuring). Therefore, faults alignment usually creates permeability contrast that is associated with groundwater emergence. Thus, the flow of groundwater along faults is either attributed to the contrast in the lithologies of the juxtaposed rocks or to the resulted crushed, permeable and porous zone formed along the fault line.

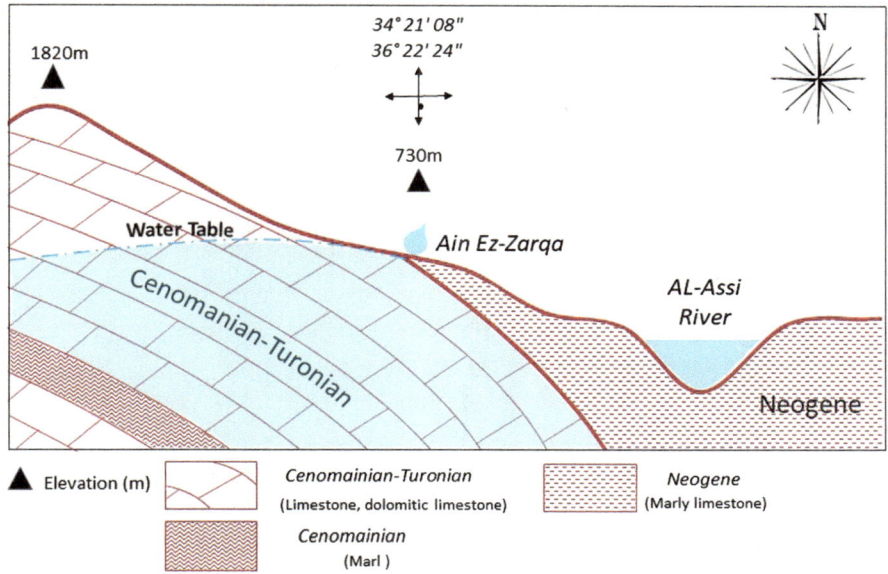

Fig. 4.2 Hydrgeologic section for Ez-Zarka Spring

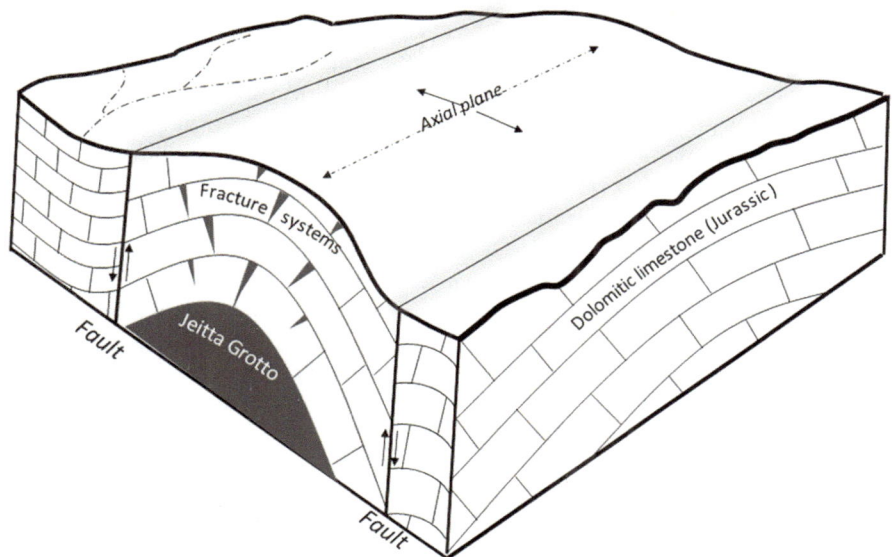

Fig. 4.3 Schematic geologic section of Jeitta Grotto Spring (Shaban 2003)

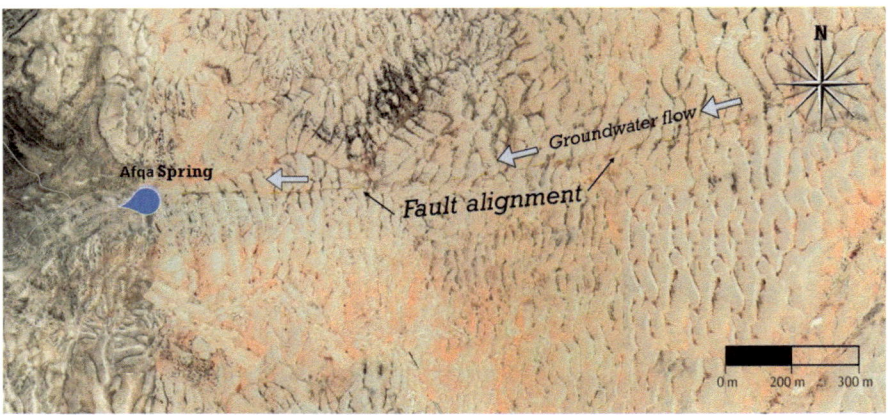

Fig. 4.4 Fault alignment transporting groundwater to Afqa Spring

This springs type is also common in Lebanon (Table 4.1), and it almost shows high discharge rate, such as springs of Rouaiss (3.55 m³/s), El-Aarbaain (3.11 m³/s) and Yammounah (2.82 m³/s).

Fault springs in Lebanon, in many instances, transport groundwater for long distances that sometimes exceed tens of kilometers, and they can be found associated with karstification, where it becomes tedious to define them as fault springs or karstic springs. This is the case of Afqa Spring, Rouaiss Spring and El-Jouz Spring; therefore, they can be attributed to both types.

Afqa Spring represents a typical example where the fault alignment can be well identified from satellite images (Fig. 4.4). The existed fault transport groundwater derived mainly from the snowmelt, and thus discharges groundwater from Afqa cave along the karistified conduit (Fig. 4.4).

4. Artesian springs: This type of springs represents the flow of groundwater spontaneously on terrain surface without pumping. The flow from this type of springs seems to defy gravity; nevertheless, it is a resultant of the pressure force occurs between two impermeable surfaces (i.e. confined aquifer).

Even though, the number of artesian springs in Lebanon is relatively few and they are characterized by moderate discharge rate (Table 4.1), yet these springs are discharging water permanently, the reason why they are considered as a major supply source of water in many of the Lebanese regions. This is well pronounced in springs of Habb (1.24 m³/s), Ras Al-Ain-Sour (0.88 m³/s) and Fawar Entelias (0.56 m³/s).

5. Thermal springs: Even though the classification of springs followed their mechanism of flow and then the related geological setting, yet thermal springs possess different concept where temperature is the main parameter. However, thermal

springs are usually described as a type of springs that are characterized by warm or hot water regardless of their hydrogeology.

It is unique of their type, thermal springs in Lebanon are few enough and the discovered ones are still with relatively low discharge, but their significance as an alternative energy source must give them concern.

Few studies done on geothermal water in Lebanon (Shaban 2010; Shaban and Khalaf-Keyrouz 2013), including thermal springs which were classified into three aspects, these are: terrestrial springs, sub-marine springs and geo-thermal water found boreholes. Most of these sources were attributed to the contact of groundwater with basalt rock masses at undefined depth. This is the reason why most of them were found in regions dominant with basalt, such as in Akkar and Ghajar regions.

Therefore, only one thermal spring (as warm water seeps) was reported. This is in Akkar Region (North Lebanon). The spring and the seeps are called "Ayoun Es-Samak" Springs where they occur at the following geographic coordinates: 34° 37′ 05″N & 35° 57′ 59″E.

The average annual discharge from this spring was estimated between 1 and 2 l/s on average. While, field measures recorded an average temperature ranging between 50 and 65 °C.

There are also a number of geo-thermal sub-marine springs and vent which were noted during several marine investigation. Thus, two major ones were found, first near Chekka coast in North Lebanon and the second is near Sour coast in South Lebanon (Shaban and Khalaf-Keyrouz 2013). Whereas El-Hage et al. (2020) reported a water plum in the sea adjacent to Akkar coast where it showed relatively warm water.

4.2.2 Springs Discharge

The number 5050 springs, as reported by MoEW and UNDP (2014), was calculated from the topographic maps which were elaborated in 1963. However, a large number of these springs is no longer exist as a result of the newly raised changes in climate and population. Therefore, about 3000 springs can be considered.

Other than the top 60 discharging springs listed in Table 4.1, the rest springs have an average discharge ranges between 2 and 4 l/s.

Therefore, the total discharge from the 60 top spring (considering the listed discharge is only for 200 days a year) is: 1165 million m³/year.

The average discharge from the rest 3000 springs will be:

$$3.5 l / s / 1000 \times 3600 \times 24 \times 365 \times 3000 = 384 \, \text{million m}^3 / \text{year}.$$

Therefore, the total discharge from springs will be, the discharge from major 60 springs + the discharge from the res 3000 ones, will be:

$$1165 + 384 = 1549 \, \text{million m}^3 \, / \, \text{year}$$

However, considering the reported discharge (according to the most available data) in Lebanon which were listed in Table 4.1, the average discharge from these springs is about 1.09 m³/s.

Generally, there are abrupt oscillations of the discharge from these springs over different years depending mainly on the volume of water derived from snow cover, plus the increased temperature rate existed. However, the highest discharge rate from these springs was found between February and March when snow melting effectively starts, while the lowest discharge is often between July and September.

Figures 4.5 and 4.6 show an example of the annual discharge from the largest two springs in Lebanon, the Afqa and Jeitta Springs.

Based on the above discussion, about the springs characteristics in Lebanon, it clear that these water resources are significant for the water budged. However, there are some remarkable characteristics for the Lebanese springs. These can be summarized as follows:

- Some springs represent the principal water feeding source for some rivers in Lebanon, such as Afqa and Rouaiss which replenish Ibrahim River.
- The carbonate rocks, specifically the rocks of Cenomanian age which are composed mainly of limestone and dolomitic limestone, represent about 70% of groundwater reservoirs for the Lebanese springs.
- Karstic springs occupy about 55% of the Lebanese springs, and then fault and contact springs occupy about 25% and 20%; respectively.
- There is clear relationship between springs type and the discharge (Edgell 1997). Thus, out of the reported 60 springs in Table 4.1 there are 50% belong the karstic and fault springs and assigned to the first order according to Meinzer Classification (1923) (Table 4.2).

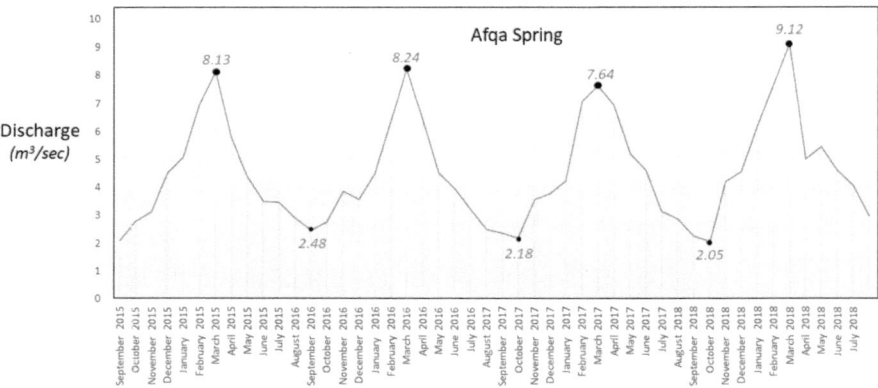

Fig. 4.5 Annual discharge from Afqa Springs

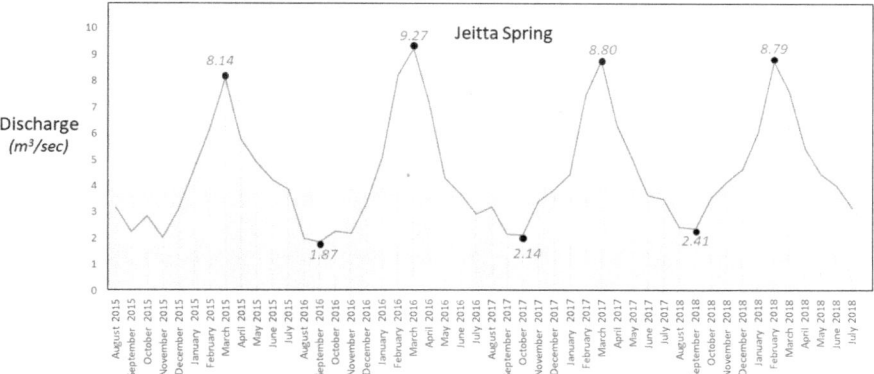

Fig. 4.6 Annual discharge from Jeitta Springs

Table 4.2 Classification of springs 'discharge (Meinzer 1923)

Magnitude	Range
First	>2.83 m³/s
Second	0.283–2.83 m³/s
Third	28.3–283 l/s
Fourth	6.31–28.3 l/s
Fifth	0.631–6.31 l/s
Sixth	63.1–631 ml/s
Seventh	7.9–63.1 ml/s
Eighth	<7.9 ml/s

4.3   Sub-marine Springs

Generally, talking about springs is often gives attention to those springs discharge water from rock masses and located in the terrestrial environment; however, it is rarely known about another type of springs that are discharging into the sea, and they called the "sub-marine" springs. The latter is a common aspect of groundwater discharge in many coastal regions worldwide. This aspect of springs is well known by fishers and people living on the coast who identify the localities of these springs into the sea where they recognize these sub-marine springs from the cooler temperature and of seawater as well as the smooth patterns appears on sea surface.

Lately with the increased water demand and scarce water resources, notably in the arid and semiarid regions; however, there has been a concern given to the nonconventional water resources. This gives the chance to think of the groundwater discharging into the sea as sub-marine springs, which became a widespread hydrologic phenomenon in many coastal zones of the world.

The discharge of groundwater into the sea occurs either at the shoreline or at a range from it. Some of these discharges are found with continuous flow all year long, while others are occur only after a rain fall peak (Shaban et al. 2005). Although of limited discharging duration the sub-marine springs are characterized with large areal extent, and they can be observed to occur over broad areas (e.g., Fielding and El-Baz 2001; Robinson et al. 2005).

Examples of sub-marine springs around the world include those off the coasts of Florida, Mexico's Yucatan Peninsula, in several areas around the Pacific Rim including Chile, Hawaii, Guam, American Samoa, and Australia, in the Arabian Gulf near Bahrain, in the Mediterranean Sea off Spain, France, Italy, Greece, Turkey, Syria, Lebanon, and Libya (SCOR/LOICZ 2004). In this respect, some obtained studies show that the discharges are directly related to open geologic structures which act to channel the water to the sea. Nevertheless, several hydrogeological controls play a role in transporting these waters from land aquifers (Shaban et al. 2005).

The Lebanese coastal line shows a number of sub-marine springs with different hydrologic characteristics. Therefore, several studies on this respect have been conducted; including also the Levantine region (El-Qareh 1967; Travaglia and Ammar 1998; CNRS-L 1999; Shaban et al. 2001, 2007, 2009a, b and 2017). Therefore, the recognition of these springs was followed by identifying their on-land sources (CNRS-L 2002; Shaban et al. 2005).

4.3.1 Concept of Identification

Water derived from land (fresh/or even waste water) is usually characterized by specific chemical, physical and even microbiological properties than those of seawater. Thus, determining these properties will help identification of water flows from land into the sea. However, applying different tests for water properties requires much work to be applied, while the presence of sub-marine springs should be primarily (and easily) recognized as a first step to be follows by more detailed steps. For this purpose, temperature is usually the key parameter used to investigate the existence of terrestrial water into the sea.

The concept behind investigation the temperature implies that terrestrial water outlets/seeps into the sea has different temperature than the seawater. Whereas, the temperature of the terrestrial water usually cooler than seawater temperature. Hence, a "thermal anomaly" is the resultant of this investigation. This is usually applied for the preliminary identification of the presence of sub-marine springs. In this respect, remarkable temperature difference is identified and then followed by applying detailed physiochemical and microbiological tests to assure of the identified water into the sea is due to freshwater (i.e. sub-marine springs) or due to wastewater resources.

In order to apply the temperature differentiation of water into the sea, there are many methods usually applied. This includes, in a broad sense, use of thermal detectors that can be directly used during the marine survey, or using remote sensing techniques, where the latter became more common since they enable surveying large areas with comprehensive terrain/coast coverage.

4.3.2   Applied Methodologies

In addition to the numerous studies done in Lebanon to locate the sub-marine springs, yet there is limited number of applied methods done to investigate these resources in-depth. However, there are two studies applied using remote sensing tools to recognize the sub-marine springs along the Lebanese coastal stretch. These are:

1. Airborne Thermal Infrared (TIR) Survey was applied by FAO (1973) for the entire Lebanese shoreline (225 km). Therefore, 99 thermal anomalies due to terrestrial water sources were determined. Later on and after applying the laboratory analysis for water sample from these sources; however, 61 of the thermal anomalies were belong to sub-marine groundwater sources.

   The estimated average discharge from the 61 sub-marine springs was about 150–200 l/s where they were distributed either directly on the coast or at distance from it. The results of this survey remained without any implementations or any exploitation plans (Shaban et al. 2005).

2. Airborne Thermal Infrared Survey was applied by CNRS-L (1997) for the northern coast of Lebanon (about 115 km). This survey showed 59 major thermal anomalies; where 27 of them were attributed to freshwater sources (i.e. submarine springs).

   For the CNRS-L TIR survey, TIR Radiometers with two bands IR-1 (2–5 μ) and IR-2 (8–14 μ) were used; in addition, TIR Radiometer with display and GPS receivers were mounted on a low-flight airplane. In this survey, the used radiometric bands could detect thermal anomalies for water flows exceeding 0.13 l/s. For each thermal anomaly, the exact coordinates were registered and thermal maps were produced (Fig. 4.7).

   Identified sites for the recognized thermal anomalies were verified in marine survey. This implied direct measuring of temperature and salinity as well as laboratory analysis to investigate the physio-chemical and microbiological properties were also applied. This helped discriminating fresh groundwater discharges into the sea from other aspects of water (e.g. wastewater outlets, etc.). Therefore, 54 sub-marine springs were recognized along the coast of Lebanon (Fig. 4.8).

4.3.3   Characteristics of Sub-marine Springs

From the produced thermal maps as well as the elaborated radiometric maps, the characteristics of the identified sub-marine springs were determined and this includes mainly their flow regime, the estimated discharge and the geologic controls behind. Therefore, the flow regime showed three major aspect as follows:

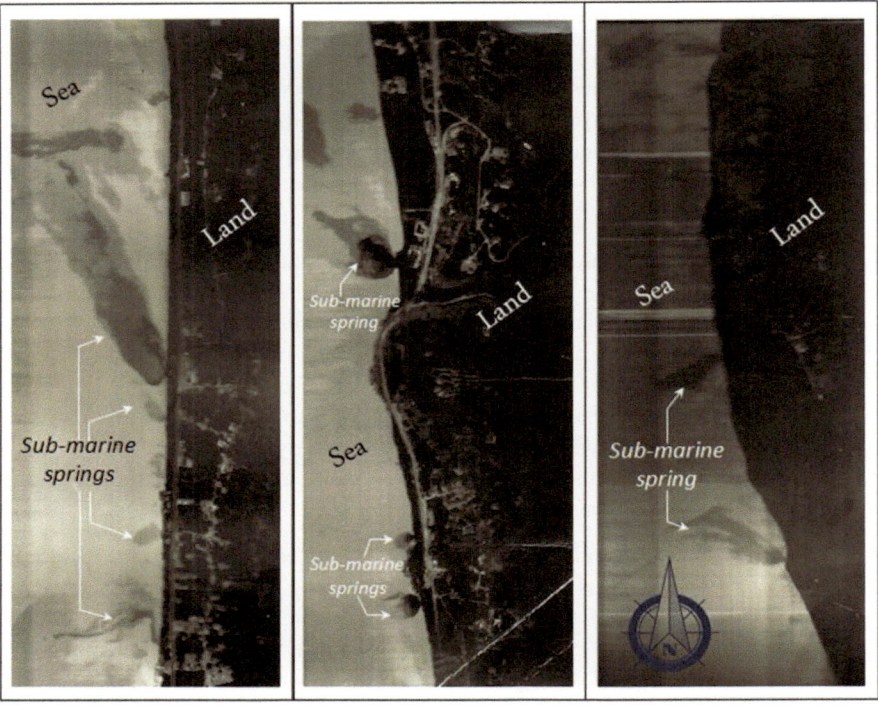

Fig. 4.7 Example for thermal maps showing the sub-marine springs along the Lebanese coast (CNRS-L 1997)

1. Perpendicular to the coastline: This is attributed to linear geologic structures, with a focus on fault alignments, karstic conduits or others features of elongated fractures.
2. Parallel to the coastline: This is attributed to non-uniform geologic structures where the flow is usually through rock tilted bed rocks (i.e. inclined rock beddings towards the sea) or fractured stratum.

Shaban et al. (2005) determined the terrestrial feeding sources for the sub-marine springs found along the northern coast of Lebanon. Thus, the applied TIR survey by CNRS-L, showed that the issuing sub-marine springs are classified as follows:

- 52% are due to karstic rock formations.
- 22% are along fault alignments.
- 18% are offshore springs.
- 67% are fed from the Carbonate rocks of the Cenomanian age.

A detailed presentation for the recognized sub-marine springs along the entire coast of Lebanon was illustrated in Table 4.3 where the results were based mainly on the TIR surveys done by CNRS-L (1997) and by FAO (1973). In addition, this was supported by field surveys applied in the terrestrial and marine environments.

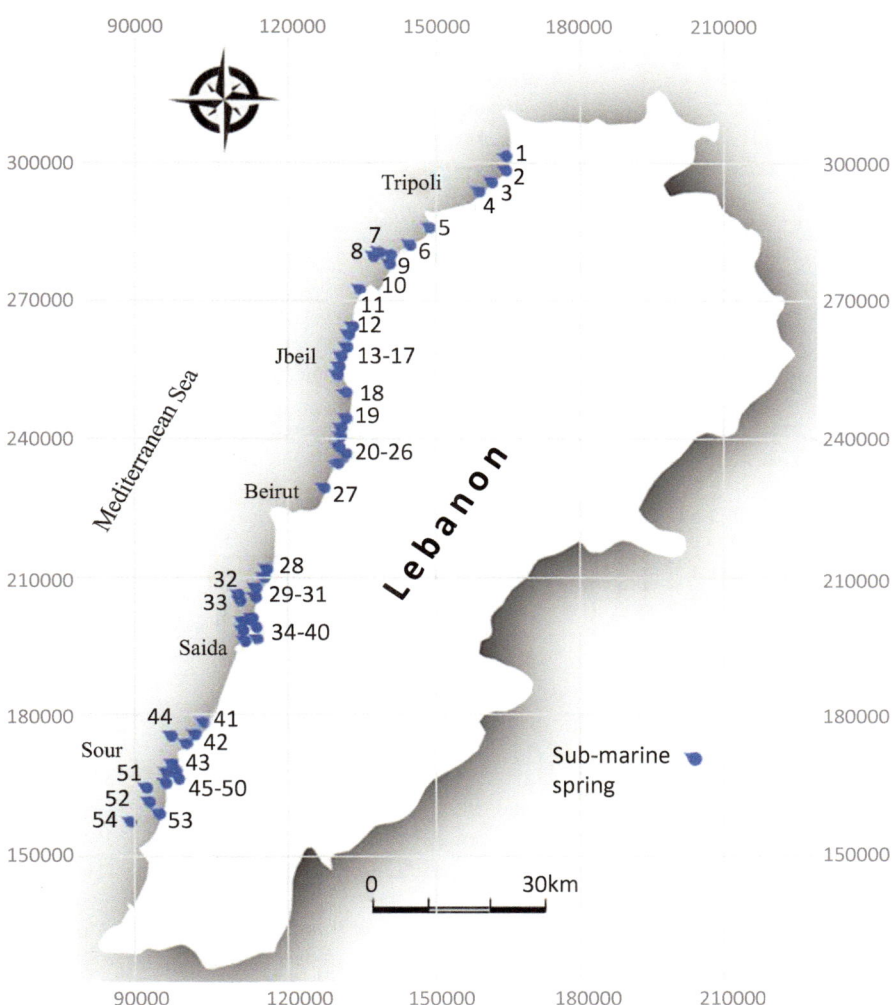

Fig. 4.8 Distribution of sub-marine springs along the coast of Lebanon

Table 4.3 Sub-marine springs along the Lebanese coast

No.[a]	Locality	Major flow regime	Discharge (*l*/s)
1	Minieh	Apparent surface water leaking from beach & parallel to it, spreading 10–15 m	60
2	Hai el-Maqateh		60
3	Bahsas	Water seeps within rocky islets extending 100 m into sea	200
4	Abu Halqa	Ditto, but extending 700 m	600
5	Chekka-1	Artesian spring 25 m off-shore, 5 m diameter water cone	500
6	Chekka-2	Ditto, 700 m off-shore, 60 m diameter	6000[b]
7	Chekka-3	Ditto, 300 m off-shore, 15 m diameter	1500
8	Chekka-4	Ditto, 300 m off-shore, 10 m diameter	1000
9	Fadous	Water seeps either parallel or protruding from rocky beach, 50–300 m	60
10	Madfoun-1	Irregular water leakage lateral & linear 40–350 m	120
11	Madfoun-2	More linear into the sea about 40–150 m	60
12	Wata el Borj	Mainly water seeps parallel, & slightly off-shore 150 m	
13	Helweh- Mar Jerjes	Mainly water seeping into the sea 300 m & slightly lateral	90
14	Mar Jerjes-1	Ditto, 100 m	60
15	Mar Jerjes-2	Ditto, 250 m	60
16	Tarol-1	Linear flow within small gulf, going 300 m off-shore	150
17	Tarol-2	Leakage parallel to beach, 100 m	60
18	Halat	Large seepage into the sea within small gulf, 800 m	500
19	Bouar-1	Linear flow into off-shore, 500 m	225
20	Bouar-2	Ditto, 300 m	120
21	Tabarja	Ditto, 600 m	250
22	Ma'ameltein-1 el-Borj	Lateral seep diffusing parallel to shore, 500 × 200 m	250
23	Ma'ameltein	Linear flow with lateral seeps, 100 × 600m	200
24	Ma'ameltein-3	Artesian spring 1 km off-shore, 30 m diameter	3000
25	Jounieh Port-1	Irregular seep from rocky beach, ≅50 m	60
26	Jounieh Port-2	Lateral flow parallel to beach, 50–100 m	60
27	Dbayeh	Linear flow within small gulf, 500 m	350
28	Khaldeh	Nearly linear flow, 200 m	60[c]
29	Doha-1	Large seepage, semi-rounded, 300 × 100m	60
30	Doha- 2	Seepage, semi-rounded, 100 × 100m	30
31	Damour-Saadyat 1	Artesian flow, 100 m off-shore, 60 m diameter water cone	60
32	Damour-Saadyat 2	Ditto, 120 m off-shore, 50 m diameter water cone	60

(continued)

Table 4.3 (continued)

No.[a]	Locality	Major flow regime	Discharge (l/s)
33	Damour-Saadyat 3	Ditto, 40 m off-shore, 25 m diameter water cone	30
34	Damour-Saadyat 4	Seepage, semi-rounded, 150 × 100m	150
35	Saadyat 1	Ditto, 75 × 90m	30
36	Saadyat 2	Elongated into the sea, 160 m	30
37	Saadyat 3	Ditto, 300 m	150
38	Oudi Ez-Ziena	Ditto, 500 m	175
39	Er-Rmayleh 1	Ditto, 400 m	175
40	Er-Rmayleh 2	Chaotic flow, 400 m	60
41	Sarafand-Aqbieh	Semi-rounded flow, 80 × 100m	30
42	Sarafand 1	Semi-elongated, 110 m	30
43	Sarafand 2	Ditto, 190 m	30
44	Khayzaran	Artesian flow 36 m off-shore, 6 m diameter water cone	30
45	Saksakieh	Semi-rounded, 130 m	100
46	Loubia	Irregular seepage, 120 m	30
47	Adloun-Nsarieh	Irregular huge flow, 500 × 900 m	500
48	Adloun	Chaotic seepage, 80 m in diameter	200
49	Ras Mienet	Semi-rounded, 40 m	30
50	Abou Al-Aswad	Chaotic flow	100
51	Ras Mienet Chaourane	Artesian flow 50 m off-shore, 10 m diameter water cone	60
52	Boroghlieh	Artesian flow 50 m off-shore, 80 m diameter water cone	200
53	Boroghlieh	Elongated into the sea, 150 m	150
54	Boroghlieh off-shore	Artesian flow 3000 m off-shore 500m diameter water cone	2000

Adapted from Shaban (2003)
[a]Numbers according to the map in Fig. 4.7
[b]Estimated by El-Qareh (1967)
[c]Estimated by FAO (1973)

It was clear that at least 46% of the recognized springs are due to karstic regime flow which is mainly along conduits and galleries. Thus, 33% of these spring are derived along elongated structures in particular the fault systems. While, the rest 21% is due to seeps from the bed rocks which are titled towards the sea (Table 4.3).

The discharge from the identified spring is different; and thus it was ranged from 30 l/s and 6000 l/s (Table 4.3). While the average is about 238 l/s. Therefore, the estimated water volume from these springs is about 410 million m^3/year, which is equivalent to the volume of water outlets from three coastal Lebanese rivers.

Chapter Highlights

- There are about 2000 springs in Lebanon with an average discharge rate of about 12 l/s.
- The total annual water volume from springs is estimated at 1140 million m^3.
- There are five major types of spring including contact, karstic, fault, artesian and thermal springs where the karstic type occupies about 55%.
- Some springs represent the principal feeding water source for some rivers in Lebanon.
- There are 54 sub-marine springs along the Lebanese coast where 15 of them are off-shore ones, and discharging at a range from the coast.

References

Abbud M, Aker N (1986) The study of the aquiferous formations of Lebanon through the chemistry of their typical springs. Lebanese Sci Bull 2(2):5–22

Azar D (2000) Contribution à l'étude de la source karstique d'Afka. Mémoire de DEA, CREEN, Université Saint-Joseph, Liban, 104p

CNRS-L (1999) Airborne thermal infrared survey to identify submarine springs along the Northern Coastline of Lebanon.. Ministry of Water Resources. Technical report, 67 pp.

CNRS-L (2002) Application of remote sensing in studying the hydrogeology of occidental Lebanon. Technical report, 52 pp.

Edgell H (1997) Karst and hydrogeology of Lebanon. Carbonates Evaporites 12:220–235

El-Hage M, Robinson C, El-Baz F, Shaban A (2020) Fracture-controlled groundwater seeps into the Mediterranean Sea along the coast of Lebanon. Arab J Geosci. Springer (Under production)

El-Qareh R (1967) The submarine springs of Chekka: exploitation of a confined aquifer discharging in the sea. Unpublished M.Sc. thesis, American University of Beirut, Geology Department, 80p

FAO (1973) Projet de développement hydro-agricole du Sud du Liban: Thermométrie aéroportée par Infra-Rouge. Programme des Nations Unies pour le developpement HG, 110, 15p

Fielding LW, El-Baz F (2001) Linear thermal anomaly offshore from Wadi Dayqah: a probable ground water seep along fracture zones, International conference on the Geology of Oman, 12–16 January 2001, Sultan Qaboos University, Muscat, Oman, Abstract vol, p 33

Guerre A (1969) Etude hydrogéologique préliminaire des karsts libanais. Hannon, Beyrouth 4:64–92

Heybrook L (1969) Introduction to hydrogeology of Lebanon. Unpublished M.Sc. thesis, American University of Beirut, Geology Department, 93p

Meinzer E (1923) Outline of groundwater hydrology with definitions: USGS Water Supply Paper, 494p

METAP (1995) Lebanon: assessment of the state of the environment. Mediterranean Environmental & Technical Assistance program. Final report. Beirut, Lebanon

MoEW (2010) Policy paper for the electricity sector, H.E Gebran Bassil Ministry of Energy and Water, June 2010

MoEW, UNDP (2014) Assessment of Groundwater resources of Lebanon, 88pp

Robinson, C.A., Buynevich, A., El-Baz, F. Shaban, A., 2005. Integrative remote sensing techniques to detect coastal fresh-water seeps. Geological Society of America Abstracts with Programs, vol 37(7):106

SCOR/LOICZ (2004) Magnitude of submarine groundwater discharge and its influence on coastal oceanographic processes. SCOR/LOICZ working Group 112. Scientific Committee on Oceanic Research. Department of Oceanography, Florida

Shaban A (2003) Studying the hydrogeology of occidental Lebanon: utilization of remote sensing. Ph.D. dissertation. Bourdeaux 1 University, 202p

Shaban A (2009a) Indicators and aspects of hydrological drought in Lebanon. Water Resour Manage J 23(2009):1875–1891

Shaban A (2009b) Monitoring freshwater discharge in the coastal zone of Lebanon using remotely sensed data. World Water Week 2009. Stockholm, Sweden

Shaban A (2010) Geothermal water in Lebanon: an alternative energy source. Low Carbon Econ 1:18–24

Shaban A (2011) Analyzing climatic and hydrologic trends in Lebanon. J Environ Sci Eng 5(3)

Shaban A (2014) Physical and anthropogenic challenges of water resources in Lebanon. J Sci Res Rep 3(3):164–179

Shaban A, Khalaf-Keyrouz L (2013) The geological controls of the geothermal groundwater sources in Lebanon. Int J Energy Environ (IJEE) 4(5):787–796

Shaban A, Khawlie M, Abdallah C (2001) New water resources for southern Lebanon: thermal infrared remote sensing of submarine springs. Conference on South Lebanon: urban challenge in the era of liberation, Beirut, 3–6/4/2001, 181–189

Shaban A, Khawlie M, Abdallah C, Faour G (2005) Geologic controls of submarine groundwater discharge: application of remote sensing to North Lebanon. Environ Geol 47(4):512–522

Shaban A, Robinson C, El-Baz F, Al-Sulaimani Z (2007) Using satellite imageries to identify groundwater channel systems along littorals of the Middle East. 4th International Conference on Wadi Hydrology. UNEP, UNESCO, RMW. Muscat, Oman, 2–4 December 2007

Shaban A, DeJong C, Al-Sulimani Z (2017) New approaches for responsible management of offshore springs in semi-arid regions. 19th EGU General Assembly, EGU2017, 23–28 April. Vienna, Austria

Travaglia C, Ammar O (1998) Groundwater exploration by satellite remote sensing in the Syrain Arab Republic. Technical report. FAO.TCP/SYR/6611, 33p

Chapter 5
Snow Cover

Abstract Among many features of water resources in Lebanon, snow is still the principal resource which has a significant role in the replenishment of rivers, springs and groundwater reservoirs. Therefore, without snow, Lebanon will lose the largest part of its water, which can be estimated to more than 60%. Recently, the exacerbated challenges on water supply makes it necessary to give concerns to snow cover and its accumulation/melting regime. This must be normal since several studies in Lebanon pointed out that snow shares to 50–60% of the water volume in rivers and springs, and then in feeding groundwater aquifers. For this reason, recent researches and studies have been applied; in particular for monitoring snow cover, field investigations, modeling and the interact with climate change. However, data analysis is still a crucial matter for research. Therefore, remote sensing techniques in combination with the advanced ground measuring stations have been utilized. This chapter will present a detailed discussion on snow cover in Lebanon depending on several research studies obtained by the author. Most of these studies utilized many types of satellite images with diverse spatial and temporal resolution. In addition, field investigations to determine snow density, depth and its relationship to different topographic and geologic features were studied. These are mostly the first of their type applied in Lebanon.

Keywords Snowpack · Sublimation · Stream flow · Water replenishments · Springs

5.1 Introduction

Snow, the solid aspect of precipitation, remains one of the best natural features that characterizes the Lebanese territory from the surrounding geographic areas in the Middle East Region. Thus, snow cover shapes the mountains of Lebanon for couple of months over the year, and then it gives a unique landscape which is reflected on

A. Shaban, *Water Resources of Lebanon*, World Water Resources 7,
https://doi.org/10.1007/978-3-030-48717-1_5

the tourism sector and the abundance of running water, plus the increasing water capacity and storage in mountain lakes, dams and ponds.

The distinguished geomorphology of Lebanon makes it different territory from the adjacent Levant regions; in particular, the existence of the two elongated mountain chains (Mount-Lebanon and Anti Lebanon) comprise remarkable meteorological ridge that obstruct the movement of cold air masses derived from the Mediterranean Sea into Eastern Mediterranean Basin. These mountainous features receive considerable amounts of snow which covers annually about one-quarter (about 2500 km²) of the Lebanese land. This is in turn reflected on the amount of water in rivers and springs, and consequently the groundwater reservoirs.

Perhaps, snowfall in Lebanon is an annual meteorological process where it can be a result of regional weather where wet storms strike Lebanon either from the European, Scandinavian or Siberian areas. Therefore, it is a common feature that snow spreads on altitude above 1500 m, and this usually remains between 4 and 8 months, and sometime, snowpack of a year is covered by snow of the consequent year. While, the melting often starts between late March. In addition, the number of snowing days is distributed as: <1 day, 2–10 days, 10–20 days, 20–40 day and >40 days for the altitudes below 250 m, 250–750 m, 750–1500 m, 1500–2250 m and >2250 m; respectively.

As a favourable hydrologic process, water derived from snowmelt regularly feeds surface and subsurface water resources unlikely to be received from rainfall. This is because the sufficient lag time needed for melting and this in turn allows uniform infiltration of water into sub-stratum. Besides, rainfall water flows rapidly (5–10 km/hour) over the steep sloping surfaces that characterize the Lebanese territory (Shaban et al. 2014). Hence, it is not an exaggeration to say that Lebanon (as country with considerable amount of water) will be a dry area without snow (Shaban and De Jong 2008).

It is agreed that water from snowmelt is the principal source of water feeding rivers, springs and groundwater. For example, it was estimated that more than 60% of water in Ibrahim River, one of the largest coastal rivers in Lebanon, is derived from snowmelt (Abd EL-Al 1953). While, an estimation declares that Mount-Lebanon receives about 1100 million m³ of water from the melting snow and this is also converted to about 425 mm/year (Shaban et al. 2004). Another example shows the significance of snow in that the snowmelt substantially feeds Jeitta Spring where the latter supplies about 75% of potable water for the Capital Beirut (Margane et al. 2013).

Recently, snowpack has been threatened and it becomes under deterioration. This is because snowpack is subjected to several physical and man-made impacts. Hence, the expansion of human activities on snow cover areas interrupted its hydrologic regime including mainly reduced percolation rate and destruction of stream flow paths along which snowmelt flows. In addition, the increased temperature with raised sunlight radiation increase the melting rate if it is compared with the past time.

Trials to evaluate the role of snow contribution in the water budget of Lebanon were put, but they were dependent only on measuring snow depth for different time periods and for selected localities from Lebanon. However, there has been a progress in the field of snow investigation and assessment, notably after the existence of water shortage. This has been raised since the beginning of 2000s where several studies have been done in Lebanon on snow cover including different hydrologic measures and this was supported by using satellite images with different spatial and temporal resolutions, as well as with the application of geo-information system (examples: Hage 2001; Shaban et al. 2004; Corbane et al. 2005; Shaban et al. 2013; Mhawej et al. 2014).

Most of these studies recommended the reservation of snowpack on the mountains of Lebanon; especially that the rock masses of these mountains represent the major aquifers of Lebanon, such as those of Jabal Akroom, Jabal Sannine and Jabal Hermoun. Hence, Shaban (2015) agreed that the elevated mountains of lebanon should be considered as natural reserves added to the existing ones in Lebanon.

Used satellite images enabled identifying the areal extent of snow cover over different time periods, even on daily basis, thus the dynamics of accumulation and melting were elaborated. This was also supported by the applications of GIS which provided good tools for the cartography of snow cover and digital data storage. This was tedious to be applied before the development of space techniques.

Yet, the ground instrumentations to measure snow characteristics are lacking in Lebanon, and all available meteorological stations are not able to measure snow depth or density. However, recently three stations have been fixed by CNRS-L in support of IRD-CESBIO (Institut de Recherche pour le Développement – Centre d'Etudes Spatiales de la Biosphère). These stations are installed in Cedars (2800 m), Laqlouq (1850 m) and El- Mazar (2300 m) (Fig. 5.1). They are able to measure, in addition to the known climatic variables, the snow depth and snow/water equivalent.

5.2   Snow Cover Area

With the development of remote sensing techniques, the geographic delineation of snow cover becomes feasible task. Therefore, satellite images are used to cartography snow cover in areas with different topographic features. Therefore, the outer border of snow cover can be calculated, but not the thickness and density of the snow pack.

5.2.1   Tools for Analysis

Data on terrain surface is acquired by using a number of space tools including aerial photographs, drones and satellite images. The latter is the most common tool used for calculating the snowpack area. It is a cheap tool and able to cover large areas whether they are remote or rugged areas.

Amongst the satellite images used by the author to calculate snow cover area, the Moderate Resolution Imaging Spectro-radiometer (MODIS) images are significant.

Fig. 5.1 Automated weather station for snow measuring in El-Mzar area, Mount-Lebanon

These images have a widespread applications in snow cover monitoring and assess-ment. They are used to study snow cover in several areas worldwide such as those done by Wang and Xie 2009; Paudel and Andersen, 2011. In some cases, MODIS images are used with other satellite images in order to gain more advantages for better images processing, such as those applied by Shaban et al. 2004 and Zhou, et al. 2013.

MODIS images are freely accessible from Goddard Space Flight Center "DISC" website of NASA Agency. The products of these images are delivered on the basis of 8-days at 250 m, 500 m and 1 km spatial resolution. These images are attributed to MODIS-Terra (MOD10-A2) and MODIS-Aqua (MYD10-A2) Snow Cover 8-Day L3 Global 500 m Grid data sets. The visibility and band ordering of these images are illustrated as (examples in Fig. 5.2):

– MODIS-Terra: true color, corrected reflectance
– MODIS-Terra: 7-2-1, corrected reflectance
– MODIS-Terra: 3-6-7, corrected reflectance
– MODIS-Aqua: true color, corrected reflectance
– MODIS-Aqua: 7-2-1, corrected reflectance
– Suomi NPP, VIIRS, corrected reflectance.

The retrieved data series were projected for Lebanon in UTM 33 North, and the different classes are merged according to "*no snow*" and "*snow and/or lake/ice*"

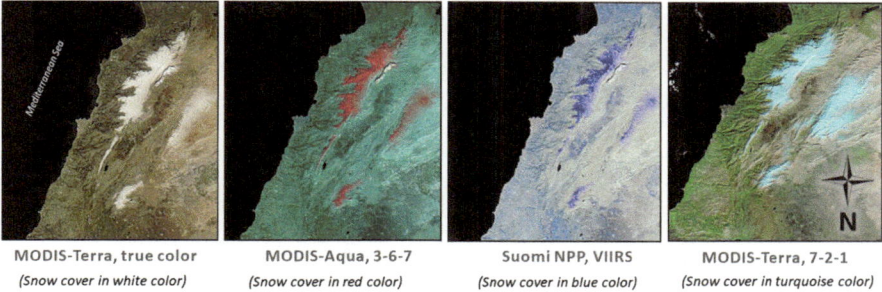

| MODIS-Terra, true color | MODIS-Aqua, 3-6-7 | Suomi NPP, VIIRS | MODIS-Terra, 7-2-1 |
| *(Snow cover in white color)* | *(Snow cover in red color)* | *(Snow cover in blue color)* | *(Snow cover in turquoise color)* |

Fig. 5.2   Example of MODIS satellite images showing snowpack

(Telesca et al. 2014). In addition, algorithms were generated to fill gaps in data from the acquired MODIS images. This was done by combining different images of Terra and Aqua; and therefore, gaps were filled with data of adjacent time periods after which a spatial majority filter was applied to assure the most precise measures for pixels adjacent to the missing measures.

In the studies obtained by the author, high resolution satellite images were also processed. In particular, Spot-4, Veg (1 km resolution) and Landsat 7 ETM$^+$ (30 m spatial resolution) were analyzed to have narrow time period observations as well as to compare snow cover area with those obtained from MODIS images (Shaban et al. 2004).

5.2.2   Data Analysis

The analysis of snow dynamics, including mainly the accumulation/melting, is significant in order to assessment many influencers and responders, such as the exited regional meteorological trends, oscillations in climatic conditions, groundwater feeding rates and the regime of stream flow. Therefore, several models were applied and aimed at creating a comparative analysis of snowpack characteristics. Similar models have been applied in many regions (examples: Serreze et al. 1999; Aouad-Rizk et al. 2005 and Martelloni et al. 2013).

The mean monthly snow cover area on the Lebanese mountain chains, were acquired from MODIS images, and then they were graphically illustrated, such as in Fig. 5.3, which shows time series from 26th February 2000 to 2nd February 2019 (i.e. over 18 years).

It is obvious that there are oscillations between the maximum and minimum measures of the snow cover area and thus peak values can be well observed (Fig. 5.3). From this figure, the average snow cover area was calculated at 737 km². Nevertheless, this value would not represent the actual snow cover area because there are several months with no snowing. Therefore, considering wet periods over all the 18 illustrated years; therefore, the average annually snow cover area was

Snow Cover Area
(km²)

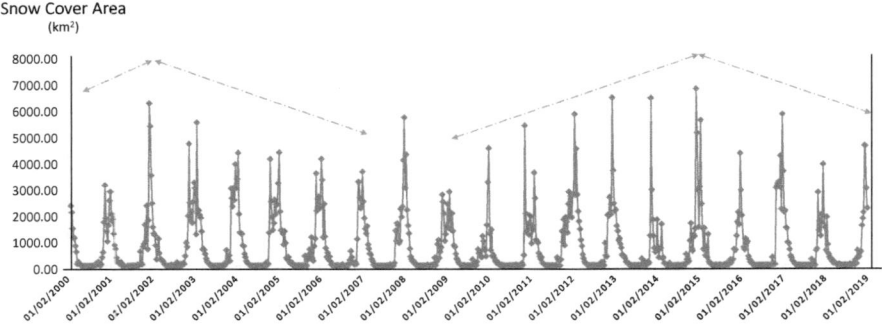

Fig. 5.3 Mean monthly snow cover area on Lebanon (2000–2018)

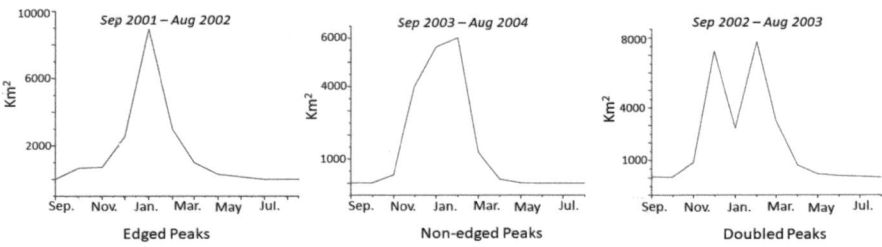

Fig. 5.4 Examples of peaks for snow cover area in Lebanon. (Telesca et al. 2014)

found as 1810 km². While the maximum reported area was 6707 km² which is equivalent to about 64% of the total area of Lebanon.

Except the reported snow cover area for the year 2008; however, a clear periodic oscillating regime existed as shown in Fig. 5.3. Hence it can be concluded that between 2000 (and may be prior), the trend is ascending until 2002; and then it was followed by descending trend from 2002 to 2007. Thus, another period occurred with ascending trend between 2009 and 2015 where it is followed with descending trend from 2015 to the beginning of 2019 and supposed to be prolonged further on (Fig. 5.3).

The illustrated time series for snow cover area in Lebanon revealed number of consecutive peaks showing the shed-points between snow accumulation and snow melt. According to Telesca et al. (2014) the existed peaks followed three different major patterns. These are:

1. Edged peaks: This is represented by sharp crests as a function of regular snow accumulation/melting processes that often occur at the same rate (Fig. 5.4).
2. Non-edged peaks: This aspect of undefined peaks indicates that the snow accumulation process is slower than the snow melting (Fig. 5.4).

3. Double peaks: When peaks occur with two or more crests (Fig. 5.4), this points out to the fact that there is immediate change in the snow cover areas that belongs to consequent meteorological storms (e.g. tropical storm, regional, etc.). Also, the newly existed seasons shifting in Lebanon can be considered as another reason for this aspect of peaks.

5.3 Water Volume from Snow

There are several methods used to calculate the volume of water derived from snow-melt (Jonas et al. 2009; López-Moreno et al. 2013; Mhawej et al. 2014; Smith et al. 2017). It is also described as snow/water equivalent (SWE), which represents water height contained in the snowpack or any the water volume captured in a defined snow volume. Thus, SWE is a function of the *water* column that would theoretically result should the whole *snow*pack melt instantaneously.

According to UNDP (2014), the average annual SWE in Lebanon was calculated between 1.82×10^6 and 2.57×10^6 million m^3 (which is equivalent to 174–246 mm of water for as measured for the years between 2008 and 2012. Besides, the estimated SWE is about 2.42×10^6 million m^3 for the years between 2002 and 2011 (Mhawej et al. 2014).

The author calculated SWE, in his studies, obtained on snow cover investigations in Lebanon. Thus, 275 different sites were selected to calculate SWE for Lebanon where the ground measures were applied almost on weekly basis. This has been done for the years between 2010 and 2013. Hence, coordinates and temperature were primarily recorded for each investigated site.

The concept behind calculating SWE implies measuring the volume of snowpack and then water density included in this snowpack. Therefore, the volume of snowpack is a function of two variables. These are the: area and thickness. Hence SWE was calculated according to the following equation (Shaban et al. 2014):

$$SWE = \left(A_{sp} \times T_{sp} \times D_{sp} \right)$$

Where:

A_{sp} is the area of snowpack.
T_{sp} is the thickness (depth) of snowpack.
D_{sp} is the density of snowpack.

Therefore, A_{sp} as the coverage extent (area) of the snow pack was calculated over the last 4 years (2010–2013) using MODIS satellite images (250 m spatial resolution). The selected data of calculated area was consistent with the same days field investigations were applied to calculate thickness and density of the snowpack.

The thickness was measured in the field either directly on the observed exposures, or by using snow coring tube especially for this purpose whenever snowpack is overburden. This was applied for each of the 275 sites, where diverse topography and altitudes were investigated.

For the snowpack density (D_{sp}), it was simply calculated by collecting snow samples in jars with identified volume, and the density was then measured after this volume of snow was diverted (melted) to water. However, several samples may be selected along one profile from the same site and when the snowpack shows diverse physical characteristics.

As a resultant, the average snowpack cover area (A_{sp}) for the investigated years was about 2585 km²; which is equal to about 25% of the Lebanese territory area. While the average snowpack thickness (T_{sp}) was 0.92 m. While the calculated (i.e. recurrence measures) snowpack density (D_{sp}) was about 73%.

Therefore, SWE for the entire Lebanon will be:

$$\text{SWE} = 2585 \times 10^6 \times 0.92 \times 73 / 100 = 1736 \quad \text{million m}^3$$

5.4 Physical Characteristics

Some studies were done on the physical characteristics of snowpack and their relationship to the physical characteristics of the site where they are located (Shook et al. 1993; Reijmer and Broeke 2003). This is significant since these characteristics are indicative for identifying the regime of snow-related processes. This can be measured directly in the field using different tools and this can be accompanied with some laboratory testis whenever it is needed.

Depending on the achieve work done by the author for the 275 sites in Lebanon, the most important physical characteristics of snowpack were investigated (Shaban et al. 2013). These are: snow density, hardness, roughness and thickness.

5.4.1 Snow Density (S_d)

Generally, the density is defined as the mass per unit volume, and it is usually expressed by kg/m³. However, if considering the volumetric measures; therefore, snow density is a function of liquid (water) volume to the snowpack volume containing this liquid. It can be also described as the percentage (%) between the volumes of water in a snowpack sample to the volume of this sample (Shaban et al. 2013).

The collection of snowpack samples, and then measuring the volume of water melted from these samples is the commonly used method to calculate the snowpack density. However, there is difference in the approaches of samples selection, and more certainly the exact points where snowpack samples must be selected. In some cases snow sample are picked from the surface of the snowpack, and this often gives erroneous results, because it is not representative to the density of the site. Some researchers consider the moderate depth to pick their sample, while others apply

Table 5.1 Classification of snowpack density

Snow density	Description	Depth[a]	Processes
<30%	Very low	Almost on surface	Old snowpack (fallen since 1 week at least), warm weather
30–60%	Low	10–20 cm	Gradational processes between the above and the below categories
60–70	Medium	20–60 cm	
70–90	Wet	60 cm – 2 m	
>90	Saturated	Deep (>2 m)	Fresh snowpack (fallen within less 1 day time)

[a]Depth was estimated according to the resulted measures done for Lebanon

Table 5.2 Classification of snowpack hardness

Snow hardness index	Description	Used object	Proposed processes
1	Very soft	Fest	Snowpack located on foot slope and avalanches
2	Soft	4 fingers	New fallen snow with regular accumulation process
3	Moderate	1 finger	Temperature- induced snow
4	Hard	Pencil	Relatively old fallen snow with relative depth
5	Very hard	Knife blade	Old fallen with high rate of melting and sublimation. It is usually on exposed surfaces.

Modified after De Quetvain (1950)

density-depth profile. The latter is more appropriate method to be applied (Takahashi and Kameda 2007).

Table 5.1 shows the empirical classification of snowpack density as adopted by Shaban et al. (2013). This classification based on the resulted measures obtained in Lebanon's territory.

5.4.2   Snow Hardness (S_h)

Hardness, as a physical property, points out to the responding of a material to any deformation processes, such as stiffness, temper, penetration, resistance to bending, scratching, abrasion, or cutting. Hence, it is commonly used the term "Ram resistance" to measure the relative hardness. This can be applied for snow where its hardness evidences many physical characteristics including mainly depth, altitude, compaction, etc.

There are many methods applied for measuring the hardness of snowpack samples with a special focus on the international classification of seasonal snow on the ground (ICSSG) where snow hardness is divided into five levels. However, "hand test" is still most widespread method where the ability to decrease an object's area is the main clue for hardness index estimation.

According to De Quetvain (1950), the most adopted ranges and descriptions for snow hardness index are shown in Table 5.2. It is obvious that the location and

Table 5.3 Snow surface roughness

Description	Roughness element	Graphical symbol	Processes
Smooth	–		Snowfall with normal condition
Wavy	Ripples		Impact of wind during snowfall
Concave furrows	Ablation, hollow, penitents		Relatively high melting and sublimation rate
Convex furrows	Rain or melt groves		Rain or melt
Random furrows	Erosion features		Impact of snow mass erosion

Modified after Kaser (2009)

timing of snowfall are significant controls for snow hardness index. Snow hardness is also a factor for melting and sublimation processes.

5.4.3   Snow Roughness (S_r)

This characteristic describes the relief of snowpack surfaces. It is also an evidence for different aspects of snow accumulation and the processes occurred later on these surfaces. In addition, the locality of snowpack is considered while studying the snow roughness, where it is influenced mainly by wind and sunlight radiation. Thus, the aspects of snow roughness and their characteristics are described in Table 5.3.

5.5   Influencing Factors

As it was discussed in Sect. 5.4, there are many studies on the physical characteristics of snowpack, but very little studies mentioned about the influencing factors. While, these factors are important since they represent the natural influencers governing snow shape and orientation. Thus, the 275 investigated sites by the author could present a comprehensive understanding about the influencing factors on snowpack characteristics.

There are six principal influencing factors affecting snowpack characterises. These are, as adopted by Shaban et al. (2013), thickness, altitude, slope, terrain property, and snow fall time and sunlight aspect (Table 5.4).

In studying snowpack thickness (i.e. depth), establishing snow depth-profile will give more accurate measuring for the overlain snow strata. Snowpack with cumulative thickness enhances compaction process as well as it works on conserving the different snow layers where the deep one occupy more water content. Therefore, five classes of thickness were adopted to characterize snowpack (Table 5.4).

Table 5.4 The controlling factors of terrain characteristics on snowpack Properties

Influencing factors	Controlling parameter[a]	Classification				
		Class-1	Class-2	Class-3	Class-4	Class-5
Thickness	Compaction	<10 cm	10–50 cm	0.5–1 m	1–2 m	>2 m
Altitude	Weather	<750 m	500-1250 m	1250-2000 m	2000-2750 m	>2750
Slope	Consolidation	<10°	10–20°	20–35°	35–45°	>45°
Terrain properties	Captured heat	Soft rocks	Partially soft	Medium	Hard	Very hard
Snowfall timing	Water retention	More than 1 month	10–30 days	5–10 days	1–5 days	Less than 1 day
Sunlight aspect	Sublimation and melting	Totally shaded	Partially shaded	Am/pm[b] exposures	Partially exposed	Exposed

Shaban et al. (2013)

[a]Parameter influences snowpack characteristics

[b]The changing of sun positioning before/after noon

Altitude of snowpack is another influencing factor on snowpack characteristics. Thus, altitude controls snow density. It is also directly proportional with snowpack cover area. Hence, five altitude classes were considered (Table 5.4). It is; therefore, well evidenced by the snow line/altitude curve adopted by (Lucas and Harrison 1990).

As an influencer on snowpack characteristics, the slope of terrain surface governs the movement and consolidation of snowpack, and it results snowpack failure (or avalanche) when it is steep, and usually a number of slope classes are accounted (Table 5.4). Hence, steep slopes tend to disintegrate snow particles, and then resulting less consolidation in the snowpack which may lead to less hardness and less density.

Other than terrain slope, there are other terrain properties which are represented mainly by the lithological characteristics. It is rarely mentioned but it was noticed by the author and et al. Hence, rock lithology underlying snowpack has a significant impact on snowpack characteristics where it plays a role the melting rate, and it implies transferring heat from the underlying rocks (Shaban et al. 2013). Therefore, consolidated rocks (e.g., limestone) capture much heat than soft and argillaceous ones (e.g., marl, clay or sand). This was obviously observed in the field where snowpack on soft rock, such as clayey remains wet for a while rather than those on hard limestone. In this view, five classes can be considered to describe terrain properties of snowpack (Table 5.4).

The time of snow fall also governs the snowpack characteristics. It is attributed the date of snow fall with respect to sampling. Therefore, newly fallen snow always reveals higher water content than old ones, and therefore, snow fallen since few days always showed much water content than those fallen since several weeks or more. For example, field observations showed that the snow fallen since a couple of days has three-times water content than that snow fallen since 1 month. For this purpose, five classes were elaborated to identify the relationship between fall time and other snowpack characteristics (Table 5.4).

Sunlight exposure is another important factor affecting snowpack, and the sunlight-exposed snowpack often s reveals different characteristics if compared with that shaded ones (snow located in shadow). Also, the field observations showed that snowpack exposed to sunlight is much hard and with less density and vice versa. This is because sunlight exposure increased the melting rate and sublimation. Thus, it can be classified in five classes as shown in Table 5.4.

5.6   Thickness-Altitude Relationship

The thickness and its relationship to altitude is usually given concern, because higher altitudes usually accompanied with thick snowpack (or depth), and this is often consistent with snow cover areas (i.e. geographic distribution of snow).

The altitude of snowpack is a function of meteorological conditions including the oscillation of atmospheric pressure, air temperature, wind, etc. this in turn governs the geographic distribution of snowpack and its physical characteristics.

The achieved field survey obtained by the author et al. on the 275 sites in the Lebanese, and the accompanied data collected at different dates on these sites, could give a comprehensive figure on the relationship between the snowpack thickness and altitude.

Therefore, the obtained results were illustrated in Fig. 5.5 where months are plotted separately to present the changing relationship between the thickness and the

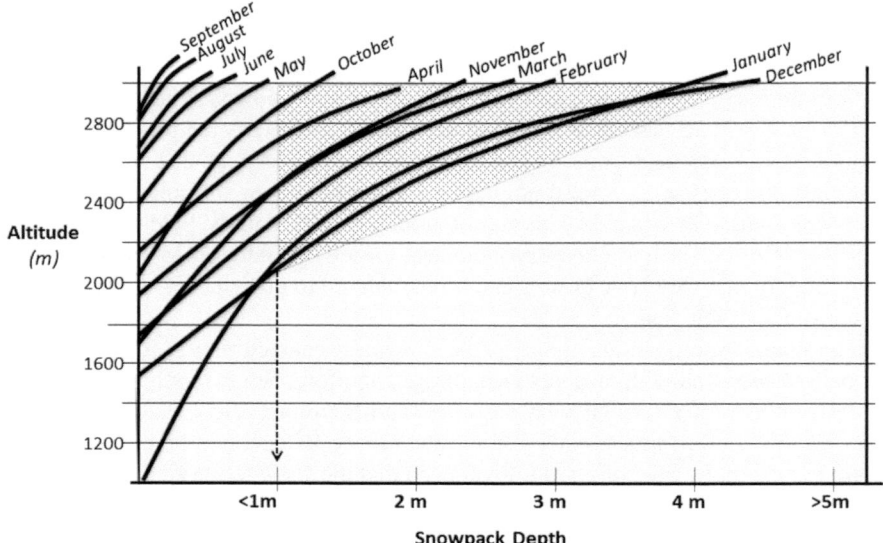

Fig. 5.5 Distribution of snowpack thickness to different altitudes in Lebanon. (Shaban et al. 2013)

altitude for each month of the year. This figure represents also all Lebanese region where snowpack remains for considerable time.

Hence, there is specific interrelationship between the snow thicknesses at a defined altitude for each month. This is based on the fact that this relationship was built on long-term data series between snowpack thicknesses according to the related altitudes. For example, snowpack with 1 m thickness often remains at altitudes of about: 2900 m, 2700 m, 2500 m, 2300 m and 2100 m for the months of October, April, November–March, February and January–December; respectively (Shaban et al. 2013).

As a matter of fact, there is no snowpack can be observed at altitudes less than 2400 m during the dry months, which extend from May to September. On the other hand, snowpack with thickness exceeding 1 m usually is observed at altitudes above 2000 m during the wet months, which extend from October to April. Moreover, snow thickness may exceed 3 m at altitudes above 2800 m during the wettest months between December and January (Fig. 5.5).

It is also clear that some months during the year show snow cover for limited time period (e.g. couple of days) which is often at moderate altitudes (i.e. less than 1100 m). Besides, other months show snow cover for a couple of months which may exceed 8–10 months, and sometimes the patches of snowpack of a define year are covered by snow of the next year.

5.7 Melting Rate

In general, the melting of snowpack is considered as a slow hydrologic process and it usually takes relatively long time period. This is the reason why snow is a major water resource in Lebanon where the snowmelt uniformly and slowly feeds the related water bodies (on surface water and groundwater resources).

There are several approaches applied to calculate the melting rate of snowpack. Thus, it can be done automatically using *in-situ* instruments (e.g. Dong et al. 2005; Egli et al. 2009, etc.). In addition, melting rate can be also measured directly in the field in order to have data series that can be statistically analysis later on. These data series are prepared for optimal calculations. For example, in the energy balance to simulate the energy fluxes or to apply the melting index approach, the most commonly used method. In addition, there is a wide use of the mathematical models to calculate the melting rate (Ross 1997).

The author applied, in his studies, measuring of snowmelt rate for different sites in the mountainous regions of Lebanon. Therefore, sinkholes were selected as the most typical localities to estimate the melting rate. This is because these karstic features are well shaped and with defined dimensions (Fig. 5.6). This in turn enables measuring the changes in the dimensions of the sinkholes, and thus the snow volume.

The dimensions, including diameter, depth and density, of snowpack in the selected sinkholes were measured at similar time intervals (i.e. 7 days). Thus, the

Fig. 5.6 Sinkhole selected to calculate the melting rate of snowpack

decreased volume of snowpack was calculated and then converted into volume of water. This can be applied after considering two variables:

1. The recharge rate to the underlying sub-stratum which was previously estimated, by Shaban and Darwich (2011), between 32 and 44%.
2. The sublimation rate which is almost negligible in the study area due to dense clouds occurred during the dates of field investigation. Nevertheless, the sublimation rate was roughly estimated at 10% after comparing it with measured values in similar regions (Bengtsson 1980; Schmidt et al. 1998; Herrero and Polo 2016).

The resulted values, from the applied field investigations, revealed that the melting rates range between 2 and 9 ml/sec, and then averaging about 6.28 ml/sec. In addition, it was evidenced that the melting rate is controlled by the altitude. For example, investigated sites above 2990 m showed a melting rate of 2.05 ml/sec, while the melting rate at 2850 m was at 9.12 ml/sec.

Whereas, snowmelt rate, according to Telesca et al. (2014), which was measured by MODIS satellite images for Lebanon showed a rate of about 16.7 *ml*/sec. This was calculated by measuring the ratio between the duration of snow accumulation and snow melting.

Chapter Highlights
- Without snow, Lebanon will lose the largest part of its water, which can be estimated to more than 60%.
- The average annually snow cover area was found as 1810 km^2.
- The annual estimated snow-water equivalent is about 1736 million m^3.
- Physical characteristics of snowpack in Lebanon have been identified including snow density, hardness and roughness.
- The controlling factors of terrain characteristics on snowpack properties where investigated including thickness, altitude, slope, terrain properties, snowfall timing and sunlight aspect.
- A detailed figure for snow thickness-altitude was illustrated. For example, snowpack with 1 m thickness often remains at altitudes of about: 2900 m, 2700 m, for the months of October, and April; respectively.
- The averaging snowmelt rate is about 6.28 ml/sec.

References

Abd EL-Al I (1953) Statics and dynamics of water in the Syro-Lebanese limestone massif. Ankara symposium on arid zone hydrology. UNESCO, Ankara, pp 60–76

Aouad-Rizk A, Job J-O, Khalil S, Touma T, Bitar C, Boqcuillon C, Najem W (2005) Snow in Lebanon: a preliminary study of snow cover over Mount Lebanon and a simple snowmelt model/Etude préliminaire du couvert neigeux et modèle de fonte des neige pour le Mont Liban. Hydrol Sci J 50(3):555–569

Bengtsson L (1980) Evaporation from a snow cover – review and dis-cussion of measurements. Nord Hydrol 11:221–234

Corbane C, Somma J, Bernier M, Fortin JP, Gauthier Y, Dedieu JP (2005) Estimation de L'équivalent en Eau du Couvert Nival en Montagne Libanaise à Partir des Images RADARSAT–1. Hydrol Sci J 50(2):355–370

De Quervain R (1950) The strength properties of the snow cover and its measurement. Geofisica purs Applicata 1950(8):13–15. (In German)

Dong J, Walker JP, Houser R (2005) Factors affecting remotely sensed snow water equivalent uncertainty. Remote Sens Environ 97:68–82

Egli L, Jonas T, Meister R (2009) Comparison of different automatic methods for estimating snow water equivalent. Cold Reg Sci Technol 57:107–115

Hage A (2001) Modélisation hydrologique conceptuelle d'un basin influencé par la couverture neigeuse. Application au Nahr El Kelb. Mémoire de DEA, CREEN, Université Saint-Joseph, Liban, 116p

Herrero J, Polo M (2016) Evapo-sublimation from the snow in the Mediterranean mountains of Sierra Nevada (Spain). Cryosphere 10(2981–2998):2016

Jonas T, Marty C, Magnusson J (2009) Estimating the snow water equivalent from snow depth measurements in the Swiss Alps. J Hydrol 378(1–2):161–167

Kaser G (2009) The International Classification of seasonal snow on ground. Technical document in Hydrology, No. 83, and IACS contribution, 2009, No. 1, 90

López-Moreno JI, Fassnacht SR, Heath JT, Musselman KN, Revuelto J, Latron J, Morán-Tejeda E, Jonas T (2013) Small scale spatial variability of snow density and depth over complex alpine terrain: implications for estimating snow water equivalent. Adv Water Resour 55:4052

Lucas R, Harrison A (1990) Snow observation by satellite: a review. Remote Sens Rev 4(3):285–348

Margane A, Schuler P, Königer P, Abi Rizk J, Stoeckl L, Raad R (2013) Hydrogeology of the Groundwater Contribution Zone of Jeita Spring. Technical Cooperation Project Protection of Jeita Spring, BGR Technical Report No. 5, 317pp. Raifoun, Lebanon

Martelloni G, Segoni S, Lagomarsino D. Fanti R. Catani F (2013) Snow accumulation/melting model (SAMM) for integrated use in regional scale landslide early warning systems. Hydrol Earth Syst Sci 17:1229–1240

Mhawej M, Faour G, Fayad A, Shaban A (2014) Towards an enhanced method to map snow cover areas and derive snow-water equivalent in Lebanon. J Hydrol 513:274–282

Paudel KP, Andersen P (2011) Monitoring snow cover variability in an agropastoral area in the Trans Himalayan region of Nepal using MODIS data with improved cloud removal methodology. Remote Sens Environ 115:1234–1246

Reijmer CH, van den Broeke MR (2003) Temporal andspatial variability of the surface mass balance in Dronning Maud land, Antarctica, as derived from automatic weather stations. J Glaciol 49(167):512–520

Ross O (1997) Mathematical models of a melting snowpack at an index plot. J Hydrol 32(1–2):139–163

Schmidt R, Troendle C, Meiman J (1998) Sublimation of snow-packs in subalpine conifer forests. Can J For Res 28:501–513

Serreze M, Clark M, Armstrong R, McGinnis D, Pulwarty R (1999) Characteristics of the western United States snowpack from snowpack telemetry (SNOTEL) data. Water Resour Res 35(7):2145–2160

Shaban A, Faour G, Khawlie M, Abdullah C (2004) Remote sensing application to estimate the volume of water in the form of snow on Mount Lebanon. Hydrol Sci J 49(4):643–653

Shaban A (2015) Application of space technology in water resources management in Lebanon. 7th World Water Forum. 12–17 April 2015, Daegu & Gyeongbuk Korea

Shaban A, Dawrich T (2011) The role of sinkholes in groundwater recharge in mountain crests of Lebanon. Environ Hydrol J 19(9):2011

Shaban A, Darwich T, El-Hage M (2013) Studying snowpack and the related terrain characteristics on Lebanon Mountain. Int J Water Sci 2(6):1–10

Shaban A, Darwich T, Drapeau L, Gascoin S (2014) Climatic induced snowpack surfaces on Lebanon's mountains. Open Hydrology Journal 2014(8):8–16

Shaban A, De Jong C (2008) Using MODIS images to characterize snow cover on the Lebanese mountains. Geophysical Research Abstracts, vol 10, EGU2008-A-11645, 2008

Shaban A, Drapeau L, Teleca L, Amacha N, Ghandour A (2020) Correlation analysis of snow cover and water capacity in the Qaraaoun reservoir, Lebanon. Water Res J

Shook K, Gray DM, Pomeroy JW (1993) Temporal variations in snowcover area during melt in prairie and alpine environments. Nordic Hydrol 24:183–198

Smith CD, Kontu A, Laffin R, Pomeroy JW (2017) An assessment of two automated snow water equivalent instruments during the WMO solid precipitation Intercomparison experiment. Cryosphere 11:101–116

Takahashi S, Kameda T (2007) Instruments and methods snow density for measuring surface mass balance using the stake method. J Glaciol 53(183):677–680

Telesca L, Shaban A, Gascoin S, Darwich T, Drapeau L, El-Hage M, Faour G (2014) Characterization of the time dynamics of monthly satellite snow cover data on mountain chains in Lebanon. J Hydrol 519(2014):3214–3222

UN Development Programme (UNDP) (2014) Assessment of groundwater resources of Lebanon. Beirut. Available at: www.lb.undp.org/content/lebanon

Wang X, Xie H (2009) New methods for studying the spatiotemporal variations of snow cover based on combination products of MODIS Terra and Aqua. J Hydrol 371:192–200

Zhou H, Aizen E, Aizen V (2013) Deriving long term snow cover extent dataset from AVHRR and MODIS data: Central Asia case study. Remote Sens Environ 136:146–162

Chapter 6
Lakes and Reservoirs

Abstract Artificial storage of surface water is a common hydrologic feature that is often observed in many arid and even humid regions. This can be also found naturally where water is accumulated by the existing terrain features. It is also widespread as an engineering implementation whereas many types of constructions are established to collect surface water. However, in both cases, this aspect of surface water resource usually contributes substantially in the water budget. Lakes and reservoirs are well known in Lebanon, where they are located in different geographic regions with remarkable existence in the mountainous ones. There are several aspects of man-made water storage which are principally governed by the topography and geology of the terrine. This water harvesting approach is adopted either on the individual or national levels. Thus, the volume of surface water storage in Lebanon has been estimated at 475 million m^3/year (Shaban, Hamzé. Shared water resources of Lebanon, Nova Publishing, New York, p 150, 2017). This is theoretically contributes to about 110 m^3/capita/year. Even though, man-made surface water storage is important for Lebanon in order to reduce surface water lose, yet there is a debate about the construction of dams. This remains a result of the lack to knowledge for the hydrological concepts. This chapter will introduce a detailed discussion and in-formation about surface water storage, artificial and man-made, in Lebanon in order to clearly provide its feasibility as a supporting water resources.

Keywords Rainfall · Surface run-off · Water harvesting · Depression-like shape · Dams · Water loss

6.1 Introduction

Generally, the term "lake" describes the natural surface water storage as it is governed by natural characteristics of terrain surface, while the term "reservoir" points out to the surface water bodies resulted from constructions made by human to store

A. Shaban, *Water Resources of Lebanon*, World Water Resources 7,
https://doi.org/10.1007/978-3-030-48717-1_6

surface water. Hence, many reservoirs are originally lakes whose discharge has been supported by constructions to control the water flow for supply purposes.

Yet, there is a misunderstanding about the terminology of lakes, reservoirs, mountain lakes, artificial lakes, etc. Nevertheless, natural surface water storage has been adopted as lakes, and when they are accompanied with man-made constructions, they become reservoir regardless of their dimension and form.

The number of lakes distributed on Earth is much bigger than the number of reservoirs. Ssatellite images and geo-information mapping technologies could identify the lakes on Earth's surface. Hence, about 117 million lakes, occupying about 4% of the World's land surface, were recognized (NG 2014).

There are about half-a-million reservoirs to the Earth's surface. This includes all the world's artificial lakes amounting to at least one hectare (10.000 m^2) in surface area. All combined, the water trapped behind dams on earth exceeds 250.000 km^2 (Downing et al. 2006).

In Lebanon, major reservoirs were made along valley courses/or on depressions. They can be characterized by larger dimensions if compared with lakes. A typical example is the Qaraaoun Reservoir which was topographically a depression-like shape that located along the primary watercourse of the Litani River. The region, where the Qaraaoun reservoir is spread, was collecting water naturally and let it on surface for long time, but large water reservoir (artificial surface water storage) has been created after the construction of the dam.

Lebanon encompasses both lakes and reservoirs where the latter are considered more reliable source of water. However, reservoirs are still not abundant in Lebanon (MoE 2005). This is derived from the concept that Lebanon is has big water volume derived from rainfall and snow in addition to the suitability of its terrain to collect surface water.

Other than the known lakes and reservoirs, mountain lakes and ground ponds are widespread water storages in Lebanon, but these constructions are limited in dimensions and abundant in number. Thus, mountain lakes become common features characterizing the mountainous landscape, while ground ponds are found as in one house out of five in the rural areas of Lebanon.

6.2 Lakes

The rugged topography of Lebanon as well as the relatively wet weather makes it common sense to presume large number of natural surface water accumulations (i.e. lakes). All these surface water bodies are found filled with water during winter and some of them retain water all year long, while others lost their water in dry seasons, and then become as wetlands.

In fact, there is no define inventory for the lakes in Lebanon, except that some large-scale ones are mentioned in many studies (CDR 2005; SNC 2011; Karrou 2014). Also, there has been a sort of conflicted nomenclatures for lakes where some of them were linked with constructions, such as dams, earth dams, channels, retaining wall, etc. Therefore, contradictory descriptions haven been raised and many of these lakes are now named as artificial lakes or reservoirs. In other words, some lakes became as reservoirs after they were undergone to constructions with different dimensions and forms.

6.2.1 Type of Lakes in Lebanon

In the view of the above discussion, naturally stored water on terrain surfaces can be diagnosed. This depended on field observations and then supported by the information acquired from satellite images. Thus, oscillating accumulations of water in these lakes at different time periods makes it tedious to put a fixed number for them. In addition, the changing precipitation rates and patterns also conflicts to make an appropriate count for the number of lakes in Lebanon.

There are two major types of lakes in Lebanon that can be described as follows:

1. Low-land lakes

These are water bodies which are accumulated on terrain surfaces as a result of the existed topography and more certainly the presence of low-land where (i.e. depression-like shapes) occur. This type of lakes has no define shape and it almost shows irregular boundary. Hence, low-land lakes are mainly fed from rainfall during wet seasons, and they partially fed from snow in the rest periods of the year; in addition, groundwater seeps contribute in feeding these lakes. Whereas many of low-land lakes become dry in summer unless they maintained by human works to isolate water from draining (Fig. 6.1).

Fig. 6.1 Low-land Lake El-Kawashra, North Lebanon

There is still a debate whether these lakes are considered as wetlands or not, but they usually encompass the characteristics of wetlands when they have considerable dimensions that almost exceed 0.05 km². Therefore, many of these lakes are wetlands in their hydrological and biological meaning.

Low-land lakes are sometime found with considerable dimensions (i.e. 200–500 m diameter); however, the largest number of them have small dimensions that averaging from tens of meters to less than hundred meter in diameter. Therefore, the average estimated water capacity of this type of lakes ranges between 200 and 5000 m³.

From the geological point of view, there is common observation of low-land lakes in that they have larger dimensions when they are located along geologic structures, notably along fault alignment or graben structures; and therefore, mixed alluviums are deposited along/or nearby the existing geologic structures, such as those located in Marjheen and Ayoun Orghosh.

The use of water from low-land lakes is still little and this can be attributed either to the intermittency of water storage in these lakes, notably in dry season when water become needed, or due to the remote sites where low-land lakes are located.

2. Excavated lakes

This type of lakes is natural in origin, and it is governed by geomorphological features that act in accumulating surface water. This type of lakes is usually found in the proximity to urban sites (i.e. almost villages in rural areas). The localities where these lakes exist include several geomorphological aspects, in particular depressions, stream meanders and springs out lets and storage terrain.

It is worth mentioning that human contribution has a major involvement in making water storage of the excavated lakes. Thus, this type of lakes is considered as a linkage between both lakes and reservoirs, and they are named by inhabitants in all Lebanese regions as "lakes" or "traditional lakes".

The main concept behind excavated lakes is that they are ancient and traditionally excavated within the villages in rural areas to collect water (Fig. 6.2), and some

Fig. 6.2 Traditional excavated lake in Maiss El-Jabal, South Lebanon

other ones have been reclaimed. Therefore, it can be estimated that between 5 and 10% of the Lebanese villages have this type of lakes.

Depending on field observation and the applied measures, it was obvious that the excavated lakes are mainly with rounded shapes where embankments are built surrounding them, and in many cases they constructed of concrete fences. While their diameter ranges between 30 and 250 m with common depth of about 10 m. therefore averaging a capacity of about 2500–5000 m^3 and sometime more.

6.2.2   Challenges on Lakes

There are many challenges have been raised lately on lakes in Lebanon. However, except the implementations taken by the inhabitants, and more certainly farmers, there are no tangible works have been given on the national level to conserve these lakes, even though demand for water has become challenging. This unfavourable situation on lakes can be attributed to several reasons as follows:

– The oscillating climatic conditions, sometimes, act in reducing water volume and then the dryness of these lakes,
– Lack to financial resources to maintain lakes, in particular to support supply methods, cleaning and other conservation approaches,
– Failure and destruction of lakes due to the existing climatic extremes and geological processes,
– Insufficient knowledge about the importance of these water bodies as non-conventional water resources,
– The chaotic urban expansion and encroachment on the feeding zones of these lakes,
– Lack of interest, by some people, for traditional methods for water supply,
– Some lakes are located in rugged and remote areas.

6.3   Reservoirs

This type of surface water bodies (i.e. reservoirs) is principally linked with the suitability of the geographic locations where they are. Therefore, these locations were prepared to store water. It is the most known aspect of surface water harvesting in Lebanon if they are compared with lakes. This is because reservoirs are established for immediate water use and they are mostly controlled and maintained.

Reservoirs are always characterized by defined shapes (e.g. rounded, elongated, one-side cut, etc.), and they are constructed according to engineering specifications where different levels and dimensions are followed. Thus, large-scale reservoirs belong/ or under the mandate of water-concerned public sector (e.g. MoEW, MoA, LRA, municipalities, etc.), and many of them have been executed under the framework of national projects (e.g. Qaraaoun Reservoir, Shabrouh Reservoir, etc.). While small-scale reservoirs are often owned by inhabitant and farmers where they use them for private purposes, notably for irrigation, and this is became common in Lebanon.

6.3.1   Dam Reservoirs

This type of reservoirs is attributed to water accumulated behind constructed dams which are done for storing water for different uses, and sometime they are considered as flood control. They are one of the most significant surface water bodies in Lebanon, in terms of dimension, after rivers and many of them are proposed from strategic point of view. They are also considered as major adaptation instruments for the changing climate and the increased water demand. That is why these reservoirs are usually given concern and many new reservoirs are being proposed.

The MoEW established a long-term plan for surface water development within the horizon of 2030. This was by proposing 18 dams, 23 lakes in the entire Lebanon, and 2 regulation weirs in the Bekaa Plain that would serve as spillways. The distribution of these dams are evenly all over the entire Lebanon. The capacities of the proposed dams vary between 4 and 128 million m3, while those of lakes vary between 0.35 and 2 million m3. This plan, if it is executed, it would allow the mobilization of an annual water volume of 1100 million m3, bringing the exploited amounts (current and future) up to 2000 million m3 (SNC 2011). Such a perspective could obviously help in providing considerable amount of water, notably for the domestic water supply and assure irrigation water for the effectively irrigable lands of Lebanon, which this around 50% of the currently cultivated lands (CDR 2005).

According to Comair (2010), the MoEW developed a 10-years plan to build dams and reservoirs in the late 1990s where 17 dams were proposed and they sought to capture approximately 650 million m3 per year. Nevertheless, this has not been executed yet, and few ones have been constructed as small-scale reservoirs, others are still as proposed and some others are under conflict, while demand for water has been exacerbated.

There are 2 large-scale dams with their reservoirs, have been established in Lebanon and 10 small-scale ones, and sometime the latter can be considered as water-retaining walls with almost limited engineering works. In addition, there are 37 proposed reservoirs where two of them are under negotiation concerning their geological setting and environmental controls (Table 6.1).

However, there are contradictory estimates about the capacity of reservoirs located behind these dams, and even those which were constructed just by to retain limited amount of water. This can be attributed to the fact that there are no enough gauges fixed to measure the capacity of some reservoirs as well as the oscillating precipitation rate in addition to the uncontrolled pumping from these reservoirs.

Even though the Qaraaoun Reservoir has an area 50 times bigger than that of Shabrouh Reservoir, as well as the capacity of the Qaraaoun Reservoir is 20 times

Table 6.1 Reservoirs and their related dams in Lebanon[a]

No	Dam name	Purpose[b]	Capacity (Mm^3)	Location[c]
Implemented Dam Reservoirs (large-scale)				
1	Qaraaoun	I & H	220	33° 34′ 03″ 35° 41′ 42″
2	Shabrouh	P	11	34° 01′ 58″ 35° 50′ 09″
Implemented Reservoirs (small-scale)				
1	Al-Bared	H & I	3.5	34° 26′ 24″ 36° 00′ 42″
2	Bnechaie	I & T	2.0	34° 20′ 02″ 35° 53′ 11″
3	Al Karn	H & I	2.0	34° 26′ 45″ 36° 06′ 50″
4	Yammounah	I & P	1.2	34° 07′ 12″ 36° 01′ 43″
5	Qamounaa	I &T	1.0	34° 29′ 18″ 36° 13′ 29″
6	Aqoura	I & P	1.0	34° 07′ 23″ 35° 53′ 50″
7	Abou Moussa	H & I	0.5	34° 28′ 42″ 35° 59′ 10″
8	El-Kawashra lake	I &T	0.35	34° 36′ 02″ 36° 12′ 16″
9	Qanan lake	H & I	0.25	33° 34′ 08″ 35° 35′ 02″
10	Wadi En-Njas	I	0.1	34° 22′ 31″ 36° 06′ 31″
Proposed Reservoirs				
1	Khardali dam	I & P	120	SL
2	Noura Et-Tahta	I	60	NL
3	Dar Beshtar	P & I	55	NL
4	Damour	I & P	40	ML
5	Qarqaf	I	30	NL
6	Al-Assi	I & P	25	B
7	Younine	I	25	B
8	Laal	I & P	10	NL
9	Mayrouba	I & P	18	ML
10	Kfar Seer lake	I	8	SL
11	Massa	I	8	B
12	Azouniyeh lake	I & P	8	ML
13	Beqa'ata	P	6.5	ML
14	Al-Mosielha	P & I	6	NL
15	Afqa	I & P	2.5	ML
16	Maasser El-Shoof	–	2	ML

(continued)

Table 6.1 (continued)

No	Dam name	Purpose[b]	Capacity (Mm³)	Location[c]
	Implemented Dam Reservoirs (large-scale)			
17	Bela'a	P	1.5	NL
18	Kififan	P	1.5	NL
19	Kfar Hounna lake	–	1.2	SL
20	Kfar Souna lake		1.1	SL
21	El Atlabi lake	–	1	NL
22	Becharre	I & P	1	ML
23	Qaisamani lake	P	1	ML
24	Rachaya lake	–	1	SL
25	Brisa	I & P	0.9	NL
26	Laqlouq lake	–	0.8	ML
27	Lebaa/Jnsnaya lake		0.8	SL
28	Wadi Sbat lake		0.7–1	B
29	Wadi Jriban lake		0.7–1	B
30	Azibeh lake		0.6	SL
31	Habash/Zaarour lake	P	0.55	ML
32	Aidamoun	I	0.3	NL
33	Tannourine	I & P	–	ML
34	El Oyoun	–	–	NL
35	Balaa lake	–	–	NL
Reservoirs with Conflict				
1	Bisri	P & I	120	33° 34′ 52″ 35° 32′ 12″
2	Janeh	P & I	38	34° 04′ 48″ 35° 49′ 53″

[a]This table was established from several sources; in addition to filed surveys done by the author
[b]H hydro-power, I irrigation, P potable, T tourism
[c]SL South Lebanon, NL north Lebanon, ML Mount Lebanon, B Bekaa

larger than that of Shabrouh Reservoir, yet both reservoirs are considered as the largest dam reservoirs in Lebanon. However, these two reservoirs have different aspects of hydrologic regimes and even diverse management approaches.

1. Qaraaoun Reservoir:

The Qaraaoun Reservoir is the largest water body in the entire Lebanon where it encompasses a surface area that changing seasonally between 8 and 12 km², and averaging storage capacity of 220 million m³. The dam was constructed in a low-land topography along the primary tributary of the Litani River where it collects water from a catchment area of about 1826 km², which is equivalent to about 17.5% of the Lebanon's surface area. The Qaraaoun Dam is characterized by 62 m height, 1090 m long and 162 m wide (Fadel et al. 2017).While, the QR encompasses an average elevation of 855 m and a maximum depth of 60 m (Fig. 6.3).

Fig. 6.3 Qaraaoun Reservoir, the largest water body in Lebanon

The reservoir, in general, is located between the major two mountain chains of Lebanon, the Mont-Lebanon and Anti-Lebanon, and more specifically in the lower part of the Bekaa Plain, where it extends between the following geographic coordinates:

33° 35′ 37″N, 33° 32′ 53″N & 35° 40′ 56″E, 35° 42′ 26″E.

These mountain chains on the two sides of the Reservoir are composed of several rock lithologies where the carbonate rock (i.e. Limestone and dolomitic limestone) are the major exposed rock lithologies. It was calculated that snow-melt contributes to about 125 million m³ through these carbonate rocks (Shaban et al. 2020). The discharge (i.e. conveying water downstream) from the Qaraaoun Reservoir is often controlled, and it ranges between 360 and 480 million m³/year, thus it averages about 420 million m³/year (Shaban and Hamzé 2018).

The main objectives of the Qaraaoun Reservoir is to irrigate about 27,500 ha mostly located downstream. In addition, it was designed to join with hydropower stations in order to generate electrical energy averaging about 500 megawatts (Shaban and Hamzé 2018).

Fig. 6.4 Shabrouh Reservoir, the second dam reservoir in Lebanon

2. Shabrouh Reservoir:

Shabrouh Reservoir is the most recent established dam reservoir in Lebanon with an area between ranges 0.16 and 0.22 km^2, and averaging storage capacity of 8–11 million m^3. The constructed dam is located along Shabrouh valley where it is surrounded by mountain ridges which are composed mainly of limestone and dolomitic limestone of the Cenomanian rock formation.

The reservoir receives water mainly from rainfall, run-off and from Laban Spring (SOER 2010) where it has a catchment area of about 18 km^2 (Fig. 6.4). In addition, the carbonate rocks provide considerable amounts of groundwater, which is originally derived from the melting snow.

Shabrouh Reservoir, which is located at 1615 m altitude, has 65 m height, 470 m long (Fig. 6.4), and it is situated between the following geographic coordinates:
34° 025′ 14″N, 34° 01′ 45″N & 34° 01′ 47″E, 35° 50′ 20″E.

The reservoir is designed to provide potable water for the downstream urban settlements where it was estimated to supply about 60,000 m^3 of potable water per day.

6.3.2 Mountain Reservoirs

Recently, mountain lakes became abundant in Lebanon; they are also called as "hill lakes" or "mountain ponds". Even though, they are mainly man-made water bodies, but they were classified as lakes due to their relatively limited water capacity as well as due to the fact the largest number of them are constructed individually by the inhabitants and framers in the mountainous regions of Lebanon.

These mountain reservoirs are constructed on different terrain surfaces, but usually at higher altitudes where precipitation is abundant, and thus they can store water from runoff/or from snowmelt to provide water for many purposes in particular for irrigation and domestic livestock as the case in Lebanon. (Shaban and Darwich 2008) reported that mountain reservoirs are usually observed at altitudes above 1500 m where snowpack often occurs.

Mountain reservoirs are designed to hold the maximum amount of water during wet months, this amount may be stored between 4 and 6 months before using it in the dry months, and this depends on the location where they were constructed.

There are many estimations about the number of mountain reservoirs in Lebanon; however, they are found to be concentrated with big numbers (as clusters) in defined regions. Hence, there are about 400 mountain reservoirs between Tannorine and Al-Aqoura (area of about 20 km^2); while there are about 120 ones between Kafer Selwan and Falogha (area of about 15 km^2).

Depending of the observations from the satellite images, there are about 2500 mountain reservoirs were estimated in the entire Lebanon. While the dimensions of these reservoirs is different, but the most commonly known capacity of these reservoirs ranges between 1500 and 5000 m^3. Build on the above numeric measures; however, the average total water capacity from mountain reservoirs can be estimated at ten million m^3.

It is well known that mountain reservoirs are increasingly demanded by inhabitants who often used them for irrigation purposes (Karrou 2014). However, the selection of the suitable sites to construct these reservoirs is always a matter of debate. This is because many of the selected sites are not potential to store water. Shaban and Darwich (2008) attributed the unsuitability of the selected sites for mountain reservoirs to the reason that many of them are constructed along stream courses where they always subjected to collapse or excessive sedimentation. While, the most recommended sites are those located along/or nearby the overflow from snowmelt which has uniform and slow flow to these reservoirs.

The shapes and design of the mountain reservoirs in Lebanon are different, but they are almost rounded where they surrounded by concrete or stony fences after covering the reservoir floor by impermeable coats or argillaceous materials.

6.3.3 Debate on Dams Construction

In many arid and semi-arid regions, considerable amounts of water supply for the domestic uses comes from reservoirs that are existed behind constructed dams. The construction of these dams is usually viewed from a national strategic vision, notably they can control flood water along streams during torrential rain periods, as well as to secure water supply during dry seasons.

Lebanon with its considerable rainfall rate and the rapid streamflow in rivers and valley systems is paradoxically does not utilize this water and the number of the constructed dams are still negligible if compared the topographic suitability for water harvesting and the increased water demand.

Nevertheless, it is unlikely that the Lebanese administration and public finance can accomplish the planned long-term plan for constructing the proposed dams and lakes before 2030. Therefore, this plan should be perceived more as a development scheme rather than a finalized and scheduled program (CDR 2005).

There are several reasons behind postpone/or obstruction of dams construction in Lebanon. The majority of this situation implies mainly the geopolitical conflicts in addition to the lack of sufficient financial resources. Nevertheless, it is a common understanding by individuals and even some decision makers that Lebanon's topography and geology are not suitable for constructing dams; in particular the seismic setting of Lebanon does not assure the stability of dams. Hence, debate on dams' construction in Lebanon often exist.

The typical examples on such debate are those on Bisri Dam and Janeh Dam. The focus on these two dams is due to the fact that the primary works have been started on both of them, but there is of postponing due the existed conflicts which are apparently attributed to their utility and geographic setting, but it can be attributed mainly to the political situation in Lebanon.

1. Bisri Dam:

    Bisri Dam has been proposed since 1953 by the American Bureau of Reclamation (USBR), and then it was adopted by the Litani Master Plan for Irrigation where the primary studies have been continued until 2000. The dam location is 35 km south of Beirut along Bisri valley close to the Bisri village.

    Few years later, CDR introduced alternative water supply plans for the Greater Beirut and Bisri Dam was on the top priorities. Therefore, implementations have been started since 2014 and it was proposed to accomplish the project by 2023.

    By the completion of Bisri Dam, water will flow through a distribution system and then providing water to 1.6 million residents in Greater Beirut and Mount Lebanon (W.B. 2019).

    The general project components can be summarized as follows:

    – Dam materials: aggregates and rubble.
    – Dimensions: height: 73 m, width: 12 m, length: 740 m.
    – Proposed lake to immerge 450 ha with water level of 460 m.
    – Storage capacity: 120 million m^3.
    – Hydro-power stations:

1. At the foot-slope of the dam with 0.2 megawatt,
2. Near Karkash station with 12 megawatt.

– Supply pipes from the dam to Joune with a total length of 3.7 km.

(a) Geologic setting:

The proposed site for Bisri Dam (which is still under construction) and the lake to be behind it, as well as the surrounding region, are located along Bisri Valley which is mostly on the upstream part of Al Awali River which fed mainly, in addition to several sources, from Naba'a El-Barook Spring.

There are several rock outcrops can be observed along the valley cut. These rocks belong mainly to the Cretaceous Epoch. Thus, the following rock formations exist (from Lower to upper):

– Neocomian-Barremian (C_1). It is composed mainly of sandstone and intervening of clay and argillaceous materials.
– Lower Aptian (C_{2a}). Clastic limestone is dominant and it is interbedded with marl and sandy limestone.
– Upper Aptian (C_{2b}). This is massive and jointed limestone and dolomitic limestone.
– Albian (C_3). It is composed mainly of marly limestone and shale.
– Cenomanian (C_4) This rock formation represents massive and well-bedded dolomitic limestone and limestone with some thin beds of marly limestone and marl.

The interbedding of the existing rocks with their diverse lithological characteristics creates surfaces of weakness between the contacts of the consequent rock formations. This separately is unfavorable geologic setting for dams' construction, but the overlying of the massive and hard dolomitic limestone of the Cenomanian age on the top makes it more compacted and rigid enough to stabilize these rocks.

There is a set of faults occur in the region. They are two major ones oriented in the NW-SE direction and span for several kilometers (Fig. 6.5). The first (4.5 km) is aligned from Bater to Gharife and the second (8 km) is from Bahnine to Bsaba Echouf.

There is also anticlinal structure which is represented by dome feature among the sandstone rocks (Neocomian-Barremian, C_1) where the axial plane of this structure extends in the NE-SW direction. Thus, the proposed location of the dam is almost at the end of the northern limb of the anticline (Fig. 6.5), and this in turn results slight terrain instability.

(b) Seismicity:

The measurements of seismic activities for the entire Lebanon has been illustrated by the Seismology Center, which belongs to CNRS-L. Therefore, instrumental seismicity has been illustrated for Lebanon where the region of the proposed Bisri Dam is included (Fig. 6.5). The available measurements are for the period between 2006 and 2017.

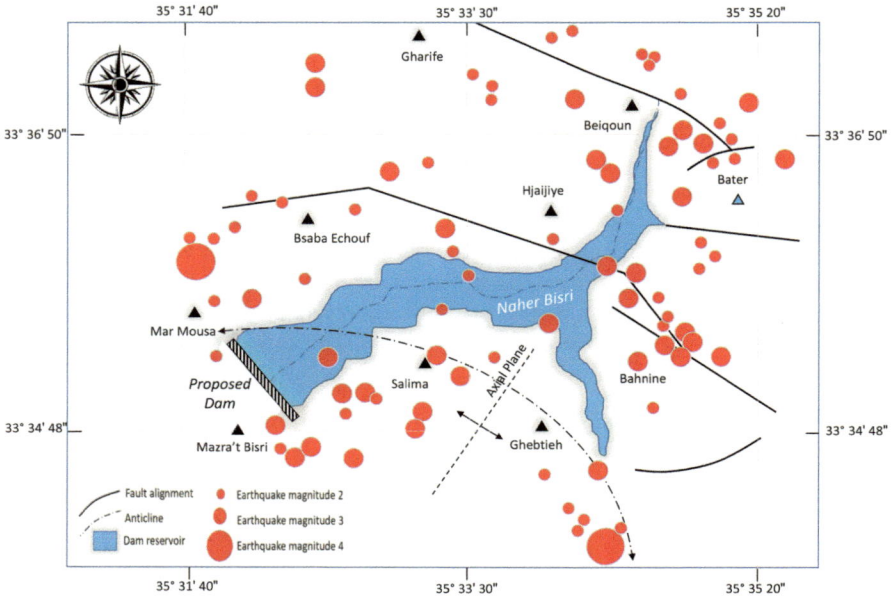

Fig. 6.5 Map showing the major geological features in the proposed Bisri Dam region

It is obvious that the frequency of earthquakes of magnitude 3 is dominant within an area of 20 km diameter surrounded the proposed dam. In addition, earthquakes with of magnitude 2 are commonly existed. However, relatively little of earthquakes of magnitude 3 occur if compared with other regions in Lebanon.

Based on the geologic setting and the existed seismic activities in the region where Bisri Dam is supposed to be built, there are some geological features act towards terrain instability, besides a moderate seismic setting for the stability of the site.

From the geological point of view, it can conclude that by considering both the geologic and seismicity in the region, there must be engineering controls to be applied as precautionary elements. This would protect the proposed dam if it is gone under any geological accident. Therefore, no effective geological constraints can be accounted for the construction of Bisri Dam.

2. Janeh Dam:

The proposed (& under construction) Janeh Dam is located along the primary watercourse of Ibrahim River, and certainly about 30 km North-East of the Capital Beirut. The studies belong to this dam have been put since 1954 after which MoEW mandated a number of consultant companies to carry on the needed assessment studies.

The dam, and its reservoir, has been designed to store about 38 million m³ of water in order to feed areas of Keserwan, Jbeil and Beirut and its suburbs.

The general project components can be summarized as follows:

– Dam materials: concrete and aggregates.
– Dimensions: height: 100 m, width: 12 m, length: 128 m.
– Water level at altitude of 735 m.
– Storage capacity: 38 million m^3 (Up to 90 million m^3)
– Hydro-power stations: A power plant with 40 megawatt will be established (CDR 2014).

(a) Geologic setting:

For the site of the proposed Janeh Dam, there is a unique rock formation exposed. It belongs to the Upper Jurassic Epoch and more certainly to Kimmeridjian age, which is composed mainly of massive and thick bedded, fractured and karistified limestone and dolomitic limestone that interbedded with thin marly horizons.

This rock formation (Kimmeridjian limestone and dolomitic lime-stone) is well characterized by rigidity and stability, notably that it has few lithological diversity, as well as its dominancy with hard and massive rocks. Therefore, it is in general a stable rock formation in terms of lithological characteristics.

In the area of concern, there are a number of large-scale and parallel faults which span for several tens of kilometers and some of them reaches the sea. Thus, one of these faults cut along the northern part of the site where the dam and its reservoir are supposed to be executed (Fig. 6.6). While there is another fault spans parallel to the previous one to the south at about 1 km distance. Besides, another parallel fault extends to the north of the area at a distance less than 3 km. Therefore, these faults are considered as negative geological features with respect to the terrain stability in the region.

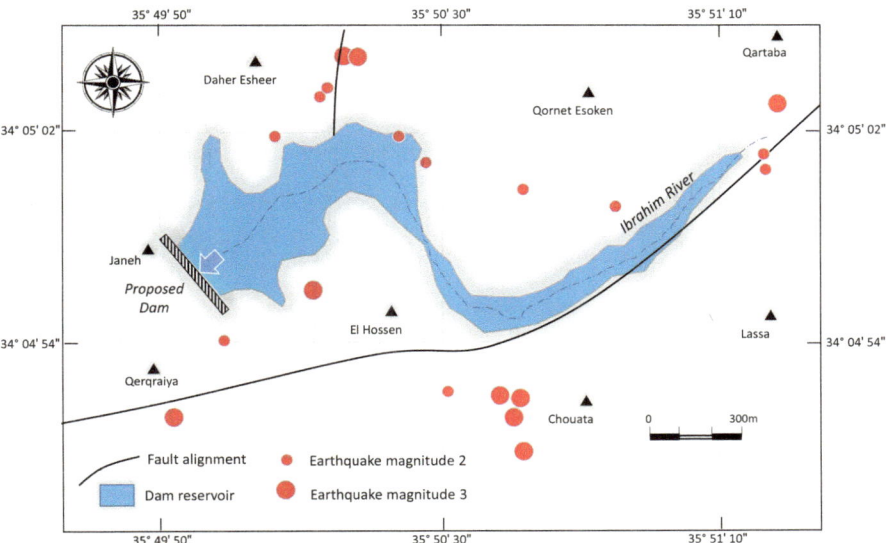

Fig. 6.6 Map showing the major geological features in the proposed Janeh Dam region

(b) Seismicity:

Likewise the case for Birsi Dam, the measurements of seismic activities for the area of concern is a part of that for the entire Lebanon which have been done by the seismology Center (CNRS-L). The instrumental seismicity measurements show little frequency and almost dominant with earthquakes of magnitude 2, while those of magnitude 3 are few, as they were illustrated for the period between 2006 and 2017. (Fig. 6.6).

As a matter of fact, the site of the proposed Janeh Dam is influenced by the geologic setting; in particular, the aligned fault systems. Nevertheless, rock lithologies are characterized by rock hardness. While, the seismicity is relatively slight according to the recorded measurements. Therefore, the site, in general, is characterized by moderate stability and engineering controls are recommended.

Chapter Highlights
- The volume of surface water storage in Lebanon has been estimated at 475 million m^3.
- There are two main natural lakes in Lebanon. These are: low-land lakes and excavated lakes where they are characterized by water capacity between 200 to 5000 m^3 and 2500 to 5000 m^3; respectively.
- There are two main dam reservoirs, the Qaraaoun Reservoir (capacity of 220 million m^3) and Shabrouh Reservoir (8–11 million m^3).
- There are about 2500 mountain reservoirs in Lebanon. The average water capacity in these reservoirs has been estimated at ten million m^3.
- There is always debate on dams' construction in Lebanon, with a focus on Bisri Dam and Janeh Dam. Based on his scientific analysis, the author agreed on the execution of these dams.

References

CDR (Council for Development and Reconstruction) (2005) National physical master plan for the Lebanese territory. CDR, Beirut

CDR (Council for Development and Reconstruction) (2014) Janneh dam construction project. Available at: http://www.ebml.gov.lb/english/project4

Comair F (2010) Water resources in lebanon. Unpublished report. MOEW to ECODIT, November, 2010

Downing J, Prairie Y, Cole J, Duarte C, Tranvik L, Striegl R, McDowell W, Kortelainen P, Caraco N, Melack J, Middelburg J (2006) The global abundance and size distribution of lakes, ponds, and impoundments. Limnol Oceanogr 51(5):2388–2397

Fadel A, Lemaire BJ, Vinçon-Leite B, Atoui A, Slim K, Tassin B (2017) On the successful use of a simplified model to simulate the succession of toxic cyanobacteria in a hypereutrophic reservoir with a highly fluctuating water level. Environ Sci Pollut Res Int:1–15. https://doi.org/10.1007/s11356-017-9723-9

Karrou M (2014) Improved water management for sustainable mountain agriculture Jordan, Lebanon and morocco Annual consultation meeting IFAD/ICARDA, Amman, 12–22/10/2014

MoE (Ministry of Environment) (2005) National Environmental Action Plan (NEAP). —Water Resources‖ (Unpublished report)

NG (National Geographic) (2014) 117 Million lakes found in latest world count. National Geographic Society Newsroom. Available at: https://blog.nationalgeographic.org

Shaban A, Darwich T (2008) Assessment of hill ponds sites in Ar-Rssal area. Unpublished technical report. Development Studies Association, 19pp

Shaban A, Hamzé M (2017) Shared water resources of Lebanon. Nova Publishing, New York, p 150

Shaban A, Hamzé M (2018) The Litani River, an assessment and current challenges, vol 85. Springer, Lebanon. https://doi.org/10.1007/978-3-319-76300-2

Shaban A, Drapeau L, Telesca L, Amacha N, Ghandour A (2020) Correlation analysis of snow cover and water capacity in the Qaraaoun Reservoir, Lebanon. Submitted to Water Resources Journal

SNC (Second National Communication) (2011) 2nd National Communication to the UNFCCC. Climate change vulnerability and adaptation. Ministry of Environment & GEF & UNDP, Beirut, 288pp

SOER (2010) The state and trends of the Lebanese environment. Ministry of Environment, UNDP. 355pp

W.B (2019) Projects: Lebanon water supply augmentation project. Available at: worldbank.org

Chapter 7
Wetlands

Abstract The occurrence of water bodies on terrain surfaces have different aspects. Thus, some of them permanently occur all year long and then they are described with define names (e.g. lake, river, etc.). However, there are, sometimes, water bodies immerse terrain surfaces intermittently and usually these bodies do not comprise a specific shape or dimension; therefore, lands where these bodies occur are described as wetlands. Lebanon, the region with relatively humid climate, encompasses a number of wetlands that spread on diverse geographic locations, and they are controlled by different hydrogeological conditions. Yet, there is no define number of wetlands in Lebanon, notably that they usually immerse terrain surfaces with irregular shapes and with different water amounts; this is in addition to the relative dimensions which often make it difficult to decide whether these are wetland or not. Moreover, the existing climatic conditions and the rapid population increase accompanied with chaotic water abstraction affected the mechanism of feeding for most of the known wetlands in Lebanon (Shaban et al. Assessment of coastal wetlands in Lebanon. In: Moran G (ed) Coastal zones: management, assessment and current challenges. Nova Science Publishers, Inc, New York, pp 27–97. ISBN:978-1-63485-611-9, 2016). There are four wetlands that were designated in the RAMSAR list. Nevertheless, the hydrogeological settings of these wetlands have not been determined yet. This chapter will present a detailed explanation on the wetlands in Lebanon and their hydrogeological interrelation; in addition to a case-study for a major wetlands in Lebanon.

Keywords Low-lands · Water level · Soil saturation · Aquifer · Ecosystem · Mediterranean

7.1 Introduction

As a unique aspect of land surface saturated with water and often characterized by green cover, "wetlands" are often mentioned in many studies and even they are accounted when calculating the water budget for many areas. This is because

A. Shaban, *Water Resources of Lebanon*, World Water Resources 7,
https://doi.org/10.1007/978-3-030-48717-1_7

wetlands became as non-conventional source of water in many regions worldwide. Thus, the RAMSAR convention counts over 2000 wetlands in the world, where 397 are found the Mediterranean Basin.

However, there are many contradictory understandings and concepts found while defining "Wetlands". This can be attributed to the fact that wetlands are viewed from different thematic aspects. Therefore, ecologists, hydrologists and other researchers explain the meaning of wetlands from their knowledge to the physical and biological elements forming these water bodies.

7.1.1 Concepts

Generally, wetlands represent irregular surface water bodies that are controlled principally by surface and groundwater feeding mechanism, but they usually exist on terrain surface for limited (& not continues) time periods. These water bodies, with their unique ecosystems, are often observed without uniform shape and dimension, but they almost comprise saturated soil all year long. According to Cowardin et al. (1992), wetlands are landforms that the saturation with water is the dominant factor determining the nature of soil development and the types of plant and animal communities.

While, Shine and De Klemm (1999) defined wetlands as dry lands saturated by water and characterized by the presence of water-dependent species of plants and animals. These, wetlands as an integral part of the water cycle, are amongst the most productive ecosystems on Earth and are of great economic and cultural importance to mankind.

Yet, there is a conflict about the natural aspects of wetlands in that many researchers consider lakes, ponds, riversides, flood plains, artificial reservoirs, etc. as wetlands, while these water bodies have their own nomenclature and description which belongs to their forming elements. Thus, wetlands should be describe only water bodies with no specific hydrologic behavior in space and time where saturated soil and s rocks with their identified flora and fauna are the major features of wetland elements.

Water-saturated lands (i.e. wetlands) are controlled mainly by water inputs into the hydrologic system, and this can be directly provided from precipitation. While, it can be indirectly or partially from other sources such as groundwater and rivers, etc. (Fig. 7.1). Thus, the geomorphological and geological factors play a role in water retention into the hydrologic system, and this in turn governs the amount of water and its continuity in supplying the wetland. Hence, the produced features of wetland occur and these are represented mainly by the biological habitat and water supply as well as the remarkable landscape (Fig. 7.1).

The territory of Lebanon, with its topography, climate, soil and geology; is suitable to create water saturation on its surface, as well as to store water on low-land surfaces, depressions and confined topography. Therefore, frequent water accumulation on these surface features is well pronounced in Lebanon where they are

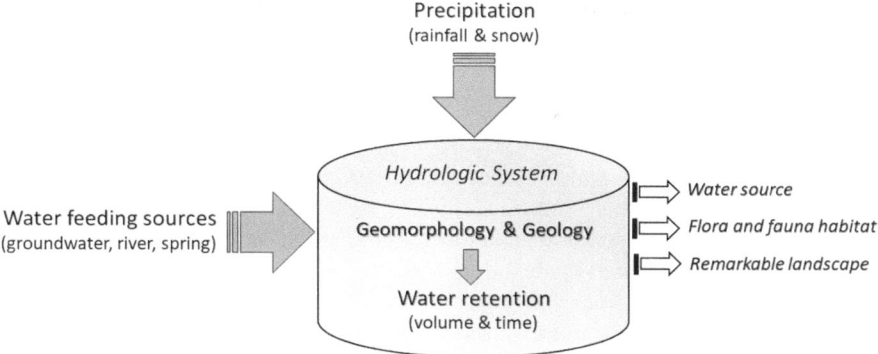

Fig. 7.1 Diagram showing the major elements of wetland formation

exposed at different scales and dimensions and then resulting unique landscape and ecosystem.

Considering the small and large-scale water bodies distributed on terrain surface; however, wetlands in the Lebanese can be considered as abundant features. Thus, most of these wetlands and saturated soil are observed in low-lands areas and depressions where rainfall often remains for a couple of months (i.e. November–April), while aquiferous rocks also occur in/or nearby the locality of wetlands.

In Lebanon, wetlands have given much concern lately when water demand has been increased. Hence, they are utilized for several purposes; in particular, for domestic water supply and irrigation and for tourism. Also, these wetlands are localities that characterized by unique species of flora and fauna. They are also identified as localities for the migratory birds fly from Africa congregate en route to Europe (Jaradi 2011). For example, Tyr Beach Wetland, in South Lebanon, occupies 204 bird species (Khater 2016). Hence, these landforms with their physical and biological components must be reserved as they constitute a significant part of the ecosystem in the region.

7.1.2   Wetlands in RAMSAR Convention

RAMSAR Convention takes a broad approach in identifying the wetlands which come under its aegis. Under the text of the Convention in Article 1.1, wetlands were described as: areas of marsh, fen, peat land or water, whether natural or artificial, permanent or temporary, with water that is static or flowing, fresh, brackish or salt, including areas of marine water the depth of which at low tide does not exceed 6 m (RAMSAR 2004).

In the above statement, it was clear that RAMSAR included the artificial aspects of surface water bodies in wetlands, as well as it also included the marine water as an environment for wetlands. This in turn has been approved by many researchers who worked on the subject matter. Nevertheless, the author does not agree on the concept that includes man-made water bodies in defining wetlands, as well as he also has no accordance on considering the marine environment as an aspect of wetlands added to the terrestrial environment. This can be attributed to the following reasons:

- The accumulation of water bodies should followed natural and spontaneous hydrologic regime that controlled by the natural influencers with a special emphasis to climatic conditions. In other words, water has to be stored on terrain surface without any man-made influencers. Nevertheless, human can help in creating enabling conditions for water storage (i.e. supporting walls, protection practices, etc.), for the naturally accumulated water on terrain surface.
- Unless it is attributed to fresh water sources, water in/nearby the marine environment is abundant as saltwater, and then it is usually distributed along the contact between the sea and the land; therefore, it cannot be assumed that the entire shoreline is a wetland. In other words, no define features can be utilized in order to discriminate the limits of a wetland along the entire marine water.

According to RAMSAR convention (2004), the interaction of several components including physical, biological and chemical ones within a wetland, such as soils, water, plants and animals, enable the wetland to perform many vital functions, in particular:

- Water storage and purification,
- Storm protection and flood mitigation,
- Stabilizing shoreline and reduce erosion,
- Groundwater recharge and discharge into/ out of the wetland,
- Retention of nutrients, sediments and pollutants,
- Acting in regulating local climatic conditions, particularly rainfall and temperature.

Also, RAMSAR convention (2004) presumed that wetlands provide several economic benefits, such as:

- Source of water, notably for domestic purposes,
- Fisheries, where 2/3 of the World's fish (especially freshwater fish) harvest is linked to the health of wetland areas,
- Agriculture, through the maintenance of water tables and nutrient retention,
- Source of energy (e.g. peat and plant matter). herbal medical products and timber,
- Recreation and tourism opportunities.
- Wildlife resources.

7.2 Wetlands in Lebanon

Lebanon, the country with sufficient water resources, is always known by water-saturated landforms. This has been reflected when RAMSAR has designated four Lebanese wetlands in the RAMSAR convention list (2004). These are the: Palm Islands, Cliffs of Ras Ech-Chekkaa, Tyr Beach and Ammiq wetlands where the last three ones are also considered as natural reserves. However, no details have been mentioned in the list about these water bodies.

In fact, there is a debate about the attributes of the known wetlands, including their hydrogeological regime, area, continuity, etc. This makes it unclear how many wetlands can be considered in Lebanon. In addition, many changes are occurring on these water bodies as a result of climatic oscillations and the increased and uncontrolled exploitation of wetland resources in Lebanon (Shaban et al. 2016).

Besides, there are many other localities which encompass the characteristics of wetlands but they have not been listed in RAMASA convention. Moreover, small-scale surface water bodies are tremendous and they can be well observed in many regions in Lebanon, but no concern has been made for them.

In this respect, it must be make it clear that the Lebanese wetlands exist with fragile ecosystems and extremely threatened by urban expansion, which have discouraged the continual development of wild life (El-Khouri 2012).

In Lebanon, concerns to wetlands remain negligible unless the MoE focused on the subject matter. However, there are also works related to wetlands have been be implemented by MoEW and MoA incorporated with many NGOs and supported by International agencies, such as World Bank, UNDP, French Global Environment Facility, etc.

Still Ammiq wetland is the most significant of its type in Lebanon where it represents typical saturated terrain. It is located north of the Qaraaoun Reservoir and it spreads on private land and it is considered one of the most important bird migration routes in the World (SOER 2010).

Tyr Beach wetland is the second wetland in Lebanon were it represents a cluster of water sources mangled with water-saturated lands. It is also a part of Tyr Beach Natural Reserve.

7.2.1 Wetlands Description

Even though RAMSA has identified only four wetlands in Lebanon, yet, there is no precise designation for the number of habitats which can be characterized as wetlands; and therefore, several localities are still considered as wetlands while they should be merely described for according to the feeding water regime, such as Jezzine Waterfall, Ed-Damour River, Yammounah Reservoir, Qaraaoun Reservoir, etc. which have been mentioned as wetlands by Wild Lebanon (2019). While they should be described as waterfall, river, and reservoirs; respectively, but not wetlands.

Nevertheless, there can be consideration for the lands neighboring to these water bodies, because they are almost saturated with water, such as that in the plunge pool at the lower end of the Jezzine Waterfall, as well as the floodplain of Ed-Damour River, and the tidal zone of Yammounah and Qaraaoun Reservoirs.

The description of wetlands can follow different natural elements. Thus, they can be classified as marine or terrestrial origin, even this is still not agreed since the marine ones (as accorded by the author) must be excluded. Moreover, wetland cannot be described according to the located fauna and flora, because the presence of a wetland controls the existence of these biological features and not the opposite.

As a primary element of wetland formation, water remains the principal acting factor. Thus, without water, wetland does not last for long time. Therefore, characterizing wetlands should be always based on water supply including (i.e. volume and continuity).

It is also worth mentioning that the geomorphology and geology of wetlands, as two main physical controls, should be always accounted. This can be characterized by comparing wetlands at different time periods and more certainly in dry and wet seasons. Figure 7.2 shows an example of where Marjheen wetland as it is observed in dry and wet season.

As per the designated wetlands in Lebanon according to RAMSA convention, perhaps there are number of in-depth parameters should be considered to designate the existence of a wetland (Shaban 2013). These are the:

1. Duration of water saturation which must be at least 3 months,
2. Spatial distribution of the saturated land where the area is significant, and it must be at least 0.05 km².

In accordance with the RAMSAR List and by considering the above parameters; however, there are ten major wetlands in Lebanon as shown in Table 7.1 and Fig. 7.3.

It is obvious that the total area of all recognized (i.e. large-scale) wetlands in Lebanon is about 15.6 km². Thus, the calculated static water volume in these water

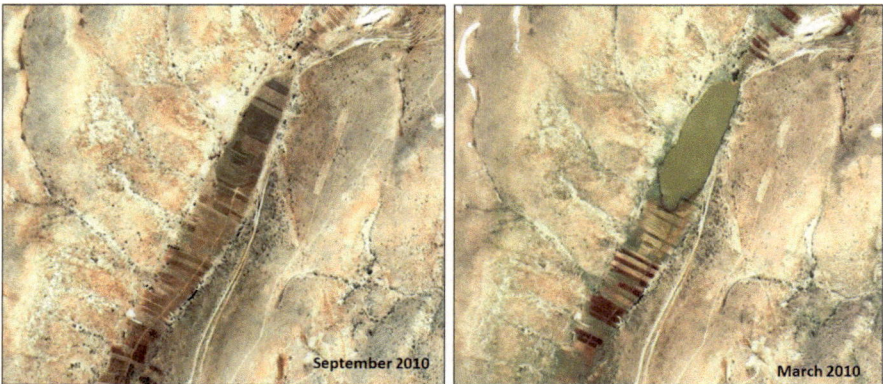

Fig. 7.2 Marjheen wetland in dry and wet periods

Table 7.1 Identified wetlands in Lebanon

No.	Name	Area (km²)	Water-feeding source	Description
1	Palm Islands	4.1	Tidal seawater	Three islands sometimes they are immersed by seawater
2	Tyr Beach	3.8	Groundwater flow along faults and bedding planes	Elongated watercourse occurs from a number of artesian springs running to the coast.
3	Aammiq	2.8	Groundwater is a major feeding source	Several natural ponds of freshwater in the carbonate rocks.
4	Aaiha	2.1	Groundwater seeps from the surrounded rock masses	Depression-like shape with abundant alluvial deposits
5	Chamsine-Anjar	0.85	Groundwater seeps and water table rise	Watercourses running water from Chamsine and Anjar spring and feed Litani River.
6	Cliffs of Ras Ech-Chakkaa	0.80	Groundwater seeps into the low-land at the sea	Jointed cliffs (~200 m) of carbonate rocks adjacent to the sea.
7	Ayoun Orghosh	0.47	Rainfall, snowmelt and groundwater seeps	Water table rise in a low-land. It is almost a confined locality
8	Marjheen	0.12	Rainfall, snowmelt and groundwater seeps	Low-land restricted between two mountain ridge
9	Kfer Zabad	0.09	A mixture of water feeding sources including rainfall and groundwater	Surface water pond and irregular water accumulation
10	Taanayel	0.07	Water table rise in low-land areas	Two pools maintained by human practices

bodies is about 80–100 million m³. Meanwhile, it can be estimated that there are about 200–250 small-scale wetlands in Lebanon where few of them are invested.

Among the designated wetlands in Lebanon, there is only one of the marine origin. It is called "Palm Islands" which have been given attention due to the unique habitat for flora and fauna rather than as a source of water.

The rest nine wetlands have different water-feeding mechanism, but it is clear that groundwater has the principal role in feeding wetlands, notably in those which are located nearby carbonate rock masses. Hence, groundwater seeps into the low-land areas is often characterized by considerable water volume that enables creating a wetland, such as that in Tyr Beach, Ammiq and Aaiha. Besides, the accumulation of water from rainfall in low-lands is usually with less volume and does not last for long time (almost less than 2 months).

7.2.2 Wetlands Degradation

Wetland degradation is the impairment of wetland functions as a result of human activity (Moser et al. 1996). Thus, degradation would be transitional phase for wetland loss which is the conversion of wetland to non-wetland areas.

Fig. 7.3 Location of the identified wetlands in Lebanon

There are many factors acting in the degradation of wetlands and these factors should be determined in order to apply the appropriate mitigation measures and controls to conserve them and let them remain as non-conventional water resources along with their remarkable biological features.

The factors of wetland degradation are different from one region to another, but they almost imply two major pillars. The natural and man-made factors, which are in many instances act together. However, when these factors act to the maximum effect, wetland loss merely occurs.

1. Natural factors:

There are many wetlands which have been subjected to major negative changes due to the physical conditions influencing their status. Hence, it is not exaggeration to say some wetlands have been totally disappeared due to the unfavorable natural conditions.

This implies, in a broad sense, the cut off (or reduce to the maximum) of water supply to feed these wetlands. It is; therefore, attributed mainly to the changing climatic conditions and its impact on water resources.

Rainfall and temperature trends are always illustrated to deduce the status of survival of wetlands, and then scenarios are often established. The changing indices are also applied to alert any threat that a wetland may suffer from. This includes the oscillating climate and the related processes. Hence, wetlands are dynamic ecosystems that are in continual change through ongoing processes of subsidence, flooding, sea level rise, drought, erosion and siltation (Shine and De Klemm (1999).

2. Man-made factors:

Unfortunately, the influence of negative human behavior on wetlands is still much more than that of the natural impact. This is attributed to the fact that human demand for water is increasing. In addition, there is also unwise use of the natural resources present among the wetlands. Thus, many wetlands have been subjected to human encroachments and the exhausting of resources.

There are many negative implements done by human on wetlands. This includes mainly:

– Chaotic pumping of water either directly from the wetland or from the sources (e.g. groundwater, springs, etc.) that feeds the wetland,
– Encroachment on wetlands by extending plantation areas, construction, etc.,
– Water contamination is a also a common phenomenon usually observed in wetlands,
– Unfavorable practices like, for example, picking some unique flora types, fishing, hunting, etc.,
– Lack to assessment approaches for the major resources in wetland (e.g. monitoring plant diseases, water table, etc.) as a result of the absence of environmental controls,
– There is ignorance to the values of heritages and archeological sites belong to the wetlands (example in Fig. 7.4).

7.2.3 Criteria to Characterize Wetlands in Lebanon

Based on the applied researches done by the author on the existing wetlands in conjunction with the RAMSAR guide points, as well as field observations and investigation the author made on different regions from Lebanon; however, a criteria to characterize these water bodies have been illustrated (UNESCO 2015; Shaban et al. 2016).

Fig. 7.4 Ignored/destroyed wetland spring in Tyr Beach Wetland

These criteria can be used as guide for better description and assessment of wetlands in Lebanon which may be also used for any other regions worldwide. Therefore, three pillars constitute the criteria to characterize wetlands as follows:

1. Water availability:

   – Existence of visible water sources

     • Rivers (when a river is found in the proximity of the wetland, it will help in the recharge process, which must be for considerable time period where it can be as overland flow or as interflow),
     • Springs (It represents the water flow from spring into the adjacent terrain),
     • Natural lake (It is the continuation of water immerse in the proximity of natural lake),

   – Existence of groundwater feeding

     • Water seepages along rock stratum (This represents the groundwater flow along bedding planes),
     • Rising peizometric level (It is the continuation of water table rise above terrain surface).

- Saturated soil

  • Level of water retention (It is viewed from the good saturation rate of the situated soil from rainfall),
  • Duration of water retention (the time that soil remains saturated).

2. Species uniqueness:

  - Unique flora

    • Species type (It is for the existence of species which are different from the surrounding regions),
    • Species number (It belongs to species number which must be exceeding the number in the surrounding regions).

  - Unique fauna

    • Species type (It is about the type of species which are different from the surrounding regions),
    • Species number (It is the number of species when it exceeds the number of species in the surrounding regions).

3. State of knowledge:

  - Reserved area (It is an area which is locally under control),
  - RAMSAR site (It belongs to wetlands as designated by RAMSAR),
  - Natural reserve (Wetland which are also considered as natural reserves),
  - Proposed area (Areas that are proposed to as wetlands).

Using the above criteria for the ten recognized wetlands in Lebanon; therefore, tables for wetland criteria assessment (Tables 7.2, 7.3, 7.4, 7.5, 7.6, 7.7, 7.8, 7.9, 7.10 and 7.11) were put to investigate the current status of the known Lebanese wetlands.

Table 7.2 Assessment criteria of Palm Islands Wetland

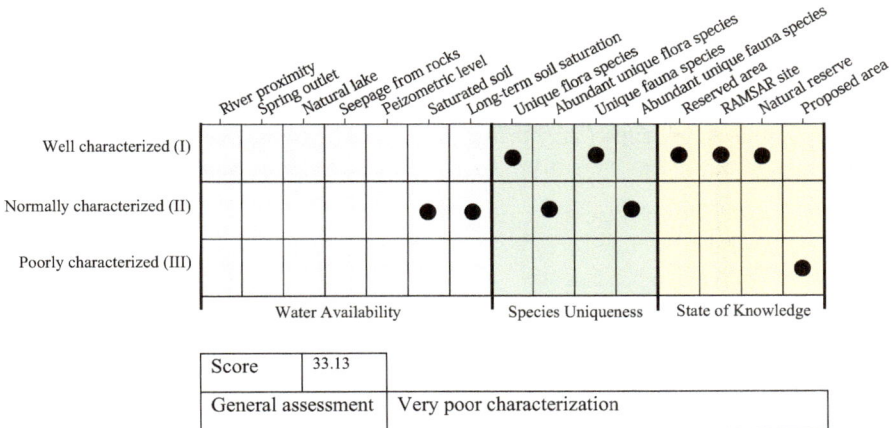

Score	33.13
General assessment	Very poor characterization

Table 7.3 Assessment criteria of Tyr Beach Wetland

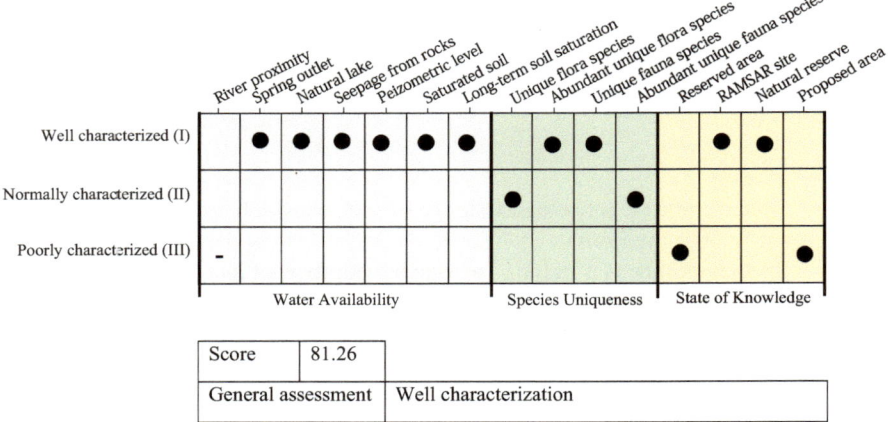

Score	81.26	
General assessment	Well characterization	

Table 7.4 Assessment criteria of Ammiq Wetland

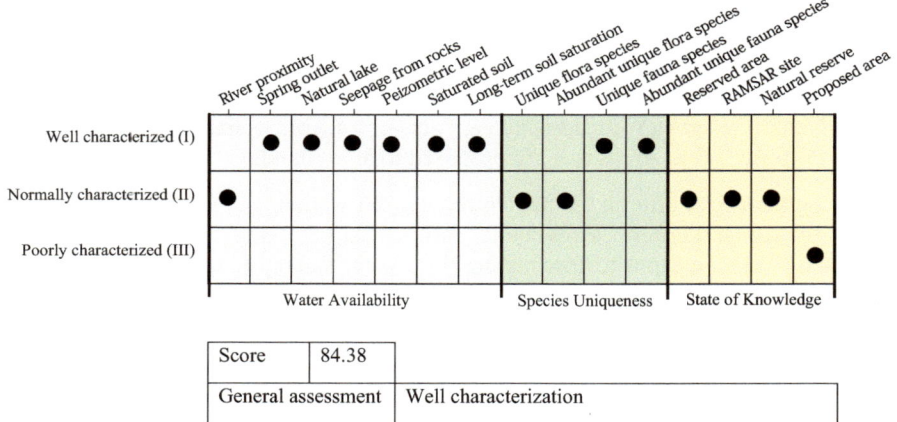

Score	84.38	
General assessment	Well characterization	

In the prepared tables, three major themes were considered with priority. These are: water availability (*primary priority*), species uniqueness (*secondary priority*) and the state of knowledge (*least priority*). According to the classification of priority, the maximum scores were given for each are as follows (Shaban et al. 2016):

– Water availability = 10 scores,
– Species uniqueness = 5 scores,
– State of knowledge = 2.5 scores.

Each of the illustrated themes (as shown in Tables 7.2, 7.3, 7.4, 7.5, 7.6, 7.7, 7.8, 7.9, 7.10 and 7.11) have three levels of description. These are: well existing, normally existing and poorly existing. Thus, the rate of each level (within a theme) will

Table 7.5 Assessment criteria of Aaiha Wetland

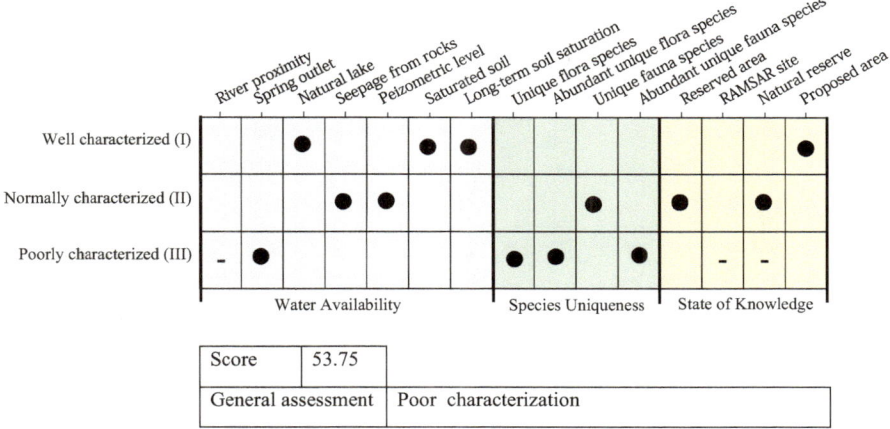

Score	53.75	
General assessment	Poor characterization	

Table 7.6 Assessment criteria of Chamsine-Anjar Wetland

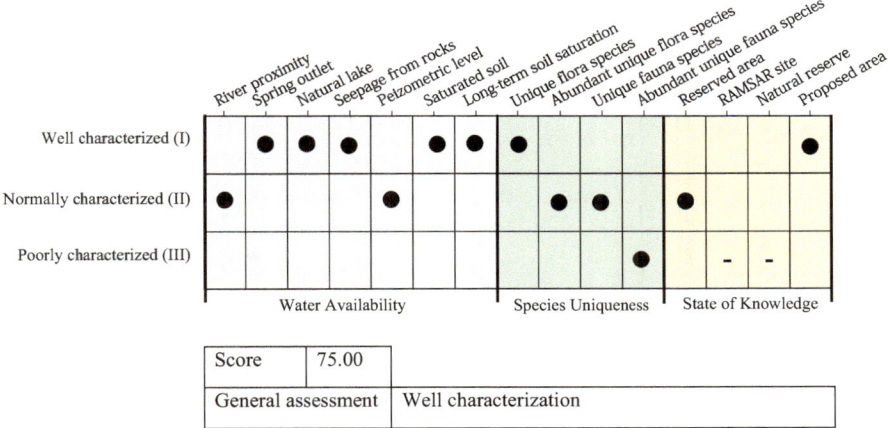

Score	75.00	
General assessment	Well characterization	

be assigned to 100%, 50% and 25% for the levels I, II and III; respectively. Therefore, the score will be as follows:

– Theme: water availability:

Level I = 10 scores
Level II = 5 scores
Level III = 2.5 scores,

– Theme: species uniqueness:

Level I = 5 scores
Level II = 2.5 scores
Level III = 1.25 scores,

Table 7.7 Assessment criteria of Cliffs of Ras Ech-Chekkaa Wetland

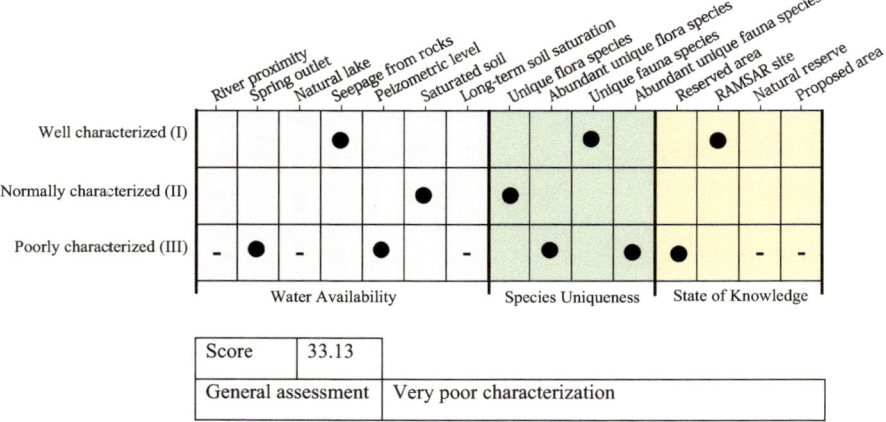

Score	33.13	
General assessment	Very poor characterization	

Table 7.8 Assessment criteria of Ayoun Orghosh Wetland

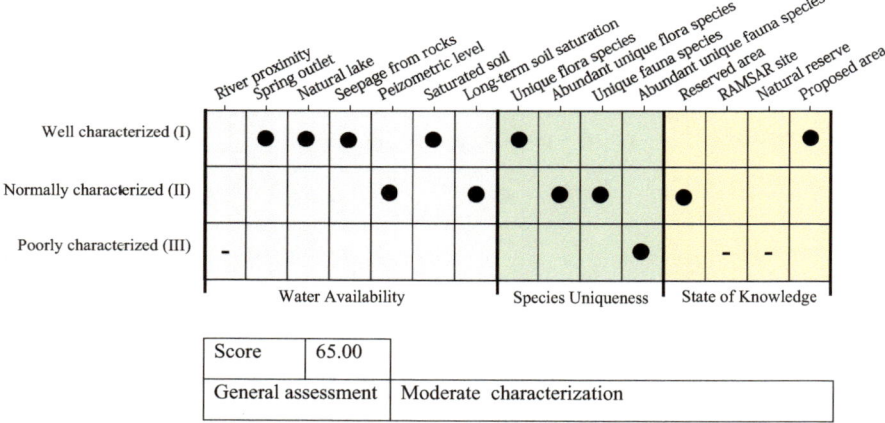

Score	65.00	
General assessment	Moderate characterization	

– Theme: state of knowledge:

Level I = 2.5 scores
Level II = 1.25 scores
Level III = 0.63 scores.

Therefore, the full cumulative scores would be 100. While, resulted values for the ten wetlands were found between 84.34% and 33.13%. Hence, the following description was elaborated to describe the wetlands based of the applied assessment:

– >80%: Very well characterization,
– 80–70%: Well characterization,

Table 7.9 Assessment criteria of Marjheen Wetland

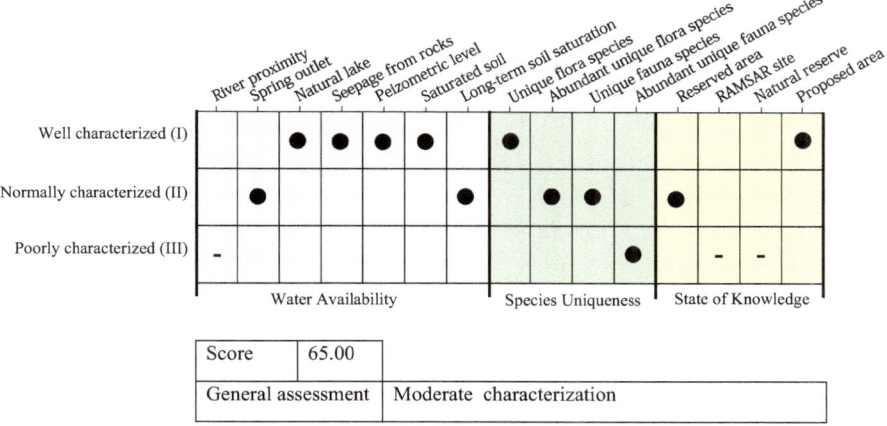

	Score	65.00
	General assessment	Moderate characterization

Table 7.10 Assessment criteria of Kfer Zabad Wetland

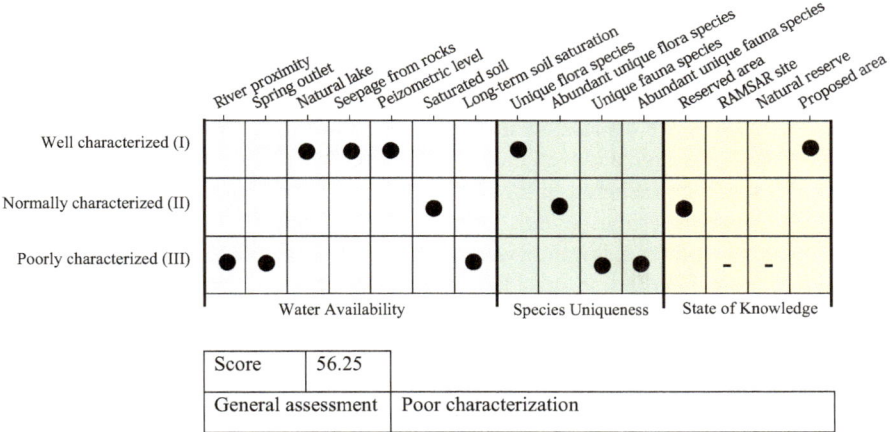

	Score	56.25
	General assessment	Poor characterization

– 70–60%: Moderate characterization,
– 60–50%: Poor characterization,
– <50%: Very poor characterization.

Tables from 7.2 to 7.11 show that there are 2 wetlands have very well characterization as they encompass the wetland characteristics. These belong to: Ammiq (84.38%) and Tyr Beach (81.20%). While, Taanayel (76.25%) and Chamsine-Anjar (75.00%) have well characterization. Consequently, Ayoun Orghosh and Marjheen were scored at 65% for each, and then they are with moderate characterization.

Table 7.11 Assessment criteria of Taanayel Wetland

	River proximity	Spring outlet	Natural lake	Seepage from rocks	Pelzometric level	Saturated soil	Long-term soil saturation	Unique flora species	Abundant unique flora species	Unique fauna species	Abundant unique fauna species	Reserved area	RAMSAR site	Natural reserve	Proposed area
Well characterized (I)	●		●		●		●	●	●			●		●	●
Normally characterized (II)		●		●		●				●					
Poorly characterized (III)											●		-	-	
	Water Availability							Species Uniqueness				State of Knowledge			

Score	76.25	
General assessment	Well characterization	

Besides, Kfer Zabad (56.25%) and Aaiha (53.75%) are poorly characterized. The rest two wetlands, which are almost related to marine environments are Palm Island and Cliff of Ras Ech-Chakkaa, were scored at 33.13% for each and then they were described as very poor criteria as wetlands.

Based on the above discussion and assessment; however, there must be concern on the existing/or designated wetlands in Lebanon. Therefore, there are some localities which should be included in the RAMSAR List for wetlands, such as those of: Taanayel, Chamsine-Anjar, Ayoun Orghosh and Marjheen. They encompass the characterstics of wetland, but they were not included in the RAMSAR List.

For Kfer Zabad and Aaiha sites, they need to be enhanced and protected in order to improve their characteristic as wetlands, as well as they need to apply projects and implementations on how to manage their water. This must be carried out by the related municipalities as well as the MoE.

For the site of Cliff of Ras Ech-Chakkaa, it was known, since a couple of years ago by saturated soil and dissipated water bodies along the coastal stretch. However, this did not last longer and the site is often dry except in the raining times. The reason behind this unfavorable situation implies the over pumping of groundwater from the neighboring mountains where groundwater feeds the wetland (UNESCO 2015; Shaban et al. 2016).

For the Palm Islands site, even though it is well characterized as a natural reserve, where it occupies abundant flora and fauna that merges between terrestrial and marine environments, yet to characterize it as wetland is not very precise from the author point of view since wetlands must be attributed merely to freshwater resources.

Chapter Highlights
- Four Lebanese wetlands have been designated by RAMSAR convention (2004); however, the author does not agreed on some of them because they do not have the physical elements to be catheterized as surface water bodies.
- Due to the increased water demand and the exploitation of the feeding water sources of wetlands in Lebanon, there is abrupt changes occurred in these wetlands.
- Small-scale surface water bodies are tremendous and they can be well observed in many regions in Lebanon. If they are well tapped, they can contribute in water supply as well as in creating unique nature.
- In addition to the wetland designated by RAMSAR, the author applied an assessment method to characterize 10 wetlands in the Lebanese territory.
- Other than the two marine wetlands, there are two wetland with poor criteria and need to be conserved.

References

Cowardin L, Carter V, Golet F, La Roe E (1992) Classification of wetlands and deep water habitats of the United States. U.S. Department of the Interior. WS/OBS-79/31. Fish and Wildlife Service, Office of Biological Services. Washington, DC, 181p

El-Khouri T (2012) Conservation Needs for Mediterranean Wetlands (with a focus on the Aammiq Wetland). Stable Institutional Structure for Protected Areas Management

Jaradi G (2011) Climate variation impact on birds of Lebanon -assessment and identification of main measures to help the birds to adapt to change. Lebanese Science Journal 12(2):25–32

Khater C (2016) The mediterranean wetlands: structure, benefits and services. Workshop on wetlands and water resources of Lebanon. UNESCO, Beirut, 12 January

Moser M, Prentice C Frazier S (1996) A global overview of wetland loss and degradation. Proceedings of Ramsar COP6 (Brisbane 1996), vol 10/12, Technical Session B at p 21

Ramsar (2004) The Ramsar Convention Manual Guide to the Convention on Wetlands Ramsar. Convention Secretariat, 75p

Shaban A (2013) Geomorphological and geological aspects of wetlands in Lebanon. 3rd International Geography Symposium (Geo-Med 2013). 10–13 June, 2013, Antalya, Turkey

Shaban A, Faour G, Stephan R, Khater C, Darwich T, Hamzé M (2016) Assessment of coastal wetlands in Lebanon. In: Moran G (ed) Coastal zones: management, assessment and current challenges. Nova Science Publishers, Inc, New York, pp 27–97. isbn:978-1-63485-611-9

Shine C, De Klemm C (1999) Wetlands, water and the Law using law to advance wetland conservation and wise use. IUCN Environmental Law Centre. Environmental Policy and Law Paper No. 38. The World Conservation Union, 340p

SOER (2010) The state and trends of the Lebanese environment. Ministry of Environment, UNDP, 355pp

UNESCO (2015) Impact of climate change on water resources of the Coastal Wetlands in Lebanon. Med-Partnership Project. UNEP, GEF, UNSECO-IHP, CNRS-L, 60p

Wild Lebanon (2019) Habitat: wetlands, lakes & rivers. Available at: http://www.wildlebanon.org/en/pages/hab/wetlands.html

Chapter 8
Groundwater

Abstract There are seventeen rock formations exposed in Lebanon. Thus, matching the precipitated water with the geological characteristics of Lebanon created typical hydrogeological sequence where permeable and porous rocks are interbedded with rock of diverse properties and then resulting a number of groundwater reservoirs and the aquiferous rock formations. There are contradictory estimations for the volume of the renewable groundwater in Lebanon. It ranges between 0.5 and 4.84 billion m³/year, besides 3.65 billion m³ in rivers and springs, but it must be clear that this groundwater volume is interrelated with those in rivers and springs and cannot be hydro-logically separated. Apart from the problem of groundwater flow into the sea as invisible rivers; yet, the groundwater resources in Lebanon are severely exhausted and the uncontrolled pumping became widespread. This added negative impact on groundwater re-sources which often found with high contamination rate. Still, the investment of groundwater does not follow scientifically-based approaches whether to explore potential reservoirs or to pump water from zones with no impact on other water resources. This chapter will present a detailed discussion on groundwater resources of Lebanon including mainly aquifer characteristics and their hydraulic properties, plus estimations for water volume and recharge rate in addition, the relationship of groundwater to faults, and then groundwater level and discharge.

Keywords Aquifer · Abstraction · Pollution · Groundwater loss · Lithological characteristics

8.1 Introduction

Usually studying groundwater needs much concern than studying other water resources. This is because groundwater, in its state, is often invisible/inaccessible, and thus hydrological and geological clues are usually investigated in order to presume the flow regime and storage of groundwater.

© The Editor(s) (if applicable) and The Author(s), under exclusive license to
Springer Nature Switzerland AG 2020
A. Shaban, *Water Resources of Lebanon*, World Water Resources 7,
https://doi.org/10.1007/978-3-030-48717-1_8

The considerable precipitation rate plus the abundant spacing systems (e.g. fractures, karst, etc.), both are accompanied with hydro-stratigraphic sequence of Lebanon, make groundwater as a linkage between the terrestrial and the marine environment which comprise an essential part of the hydrologic cycle, and thus groundwater in Lebanon is a renewable resource.

Likewise surface water resources, groundwater in Lebanon is widely tapped, and the aquiferous rock formations are subjected to intensive exhaustion (Shaban and Hamzé 2017). Therefore, large number of drilled wells are spread on different geographic regions in Lebanon, notably the agricultural lands and in the highly urbanized zones. Hence, most of these wells are private and they can be as much as 20 times the number of public ones. This in turn results unfavorable status on groundwater resources, and thus abrupt depletion in water table has been occurred. This has been reflected on depth to water, pumping rate, discharge from the surrounding springs as well as the intrusion of seawater into the coastal aquifers.

Contamination is another problem occurred in groundwater resources in Lebanon. It implies a variety of aspects including the physiochemical and biological contamination, which is a dominant phenomenon and represents a geo-environmental problem in many regions; especially in agricultural areas where shallow aquiferous rock formations (less than few tens of meters) exist. Also, contamination is well pronounced in the dense urbanized areas where tremendous liquid wastes infiltrate into the underlying substratum.

In fact, the situation of groundwater in Lebanon is unsatisfactory, and it can be said: it is degraded, while the controls put to enhance the existing situation are not enough or in many instances inappropriate to address the consequent problems of quantity, quality and even the sustainability of supply.

If compared with studies done on surface water resources, studies on groundwater in Lebanon are few and rarely applied in details except some elaborated ones by international entities incorporated as projects, such as: UNDP (1970), MoEW and UNDP (2014).

It is not exaggeration to say that, except the basics, the hydro-geology of Lebanon is still undetermined and it is being rapidly changing notably with the increased water demand and the oscillating climatic regime. The weakness in finding much studies and understandable concepts on groundwater hydrology in Lebanon can be attributed to many reasons including mainly the lack to knowledge on the subjects of Hydrogeology and Earth Sciences as a whole.

Concerns are often oriented towards surface water including number of projects and adaptation measures are usually proposed. Nevertheless, there are no creditable works or proposed projects to invest/or conserve groundwater resources in Lebanon. However, as a unique theme, the artificial recharge in the coastal aquifers is always mentioned, but hasn't applied yet.

In this respect, MoEW (2010) created the National Water Sector Strategy (NWSS), which was developed with the participation of national stakeholders and international donors. The baselines of NWSS included the theme on water sector infrastructure. Among this theme and under the title of production, there was one item mentioned the investment of groundwater and presumed digging boreholes without and specific characterization for the depth and pumping rate or geography of this investment. This reflects the lack for comprehensive knowledge on how to treat these resources properly.

8.2 Hydrogeological Characteristics

Diverse lithological properties are characterizing the rock sequence in Lebanon where the carbonate facies occupy the largest part and it is, in general, interbedded with marly and clastic facies. This geo-logical sequence is stratifying different rock formations and units, as well as beds and even lamina. These rock aspects have contradictory behavior towards groundwater flow, storage and flow retard according to the physical properties of each aspect.

In this view, the term "Aquiferous" can be used to describe the capability of rocks or rock masses to occupy groundwater or control-ling the flow of groundwater within rocks. Therefore, the following aquiferous properties represent the principal hydrogeological characteristics of rock sequence in Lebanon:

- Aquifer: A stratum, rock unit or rock formation that stores groundwater with considerable amounts and can be tapped successfully.
- Semi-aquifer: This is undefined rock in terms of groundwater storage, and it can be an aquifer, but it is often not unified.
- Aquiclude: It represents a rock (rock mass, unit, etc.) that is characterized by low permeability, and it often creates a hydrologic barrier and does not allow groundwater flow.
- Aquitard: This rock (rock mass, unit, etc.) retards the flow of groundwater, but it does not totally prevent the flow of groundwater between different rock layers.

8.2.1 Aquiferous Properties

Considering the above four aspects of aquiferous rock properties, thus the rock sequence in Lebanon can be classified accordingly (Fig. 8.1). This classification depends on the lithological characteristics of the exposed in Lebanon where the old-est rocks are attributed to the Middles Jurassic. The classification of aquiferous rocks properties (from bottom to top) can be summarized in Table 8.1.

It is obvious that the geological sequence of Lebanon occupies a variety of aquiferous rock properties where water-bearing and transit rock formations are interbedded with retarding-flow rocks and with hydrologic barriers. Thus, they can be described as follows:

1. Excellent aquifer: There are two rock formations described as "Excellent" Aquifers. This description has been adopted due to the fact that these two aquifers are retaining considerable amount of groundwater. These belong to the

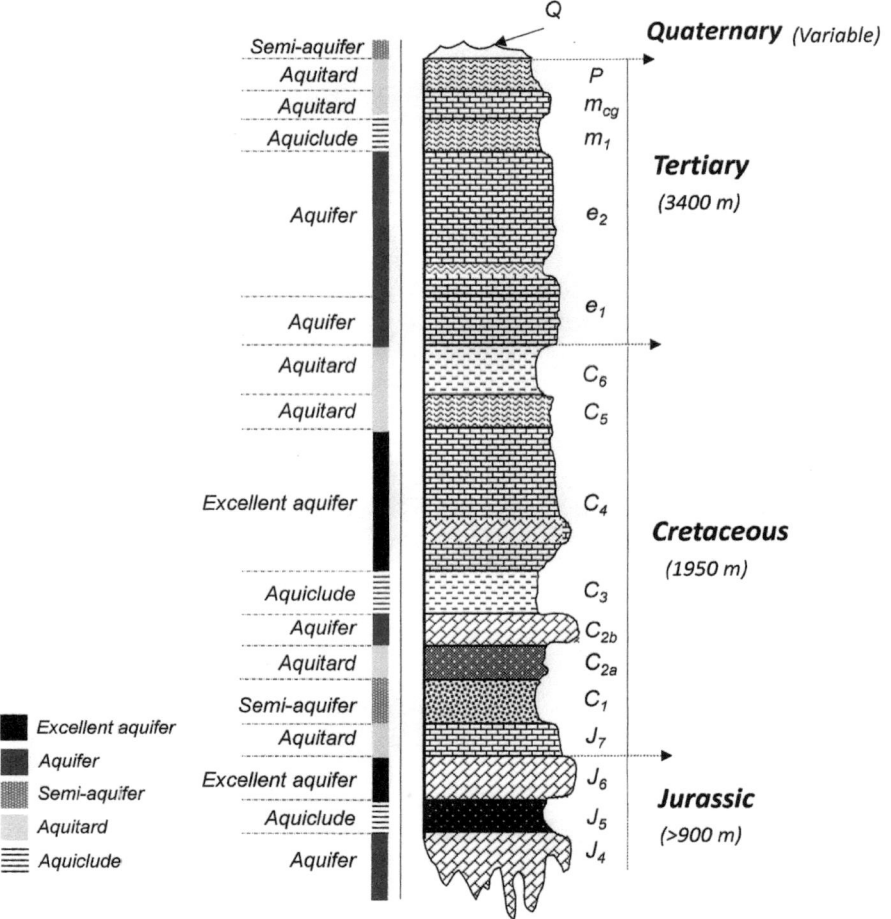

Fig. 8.1 Classification of aquiferous rock formations in Lebanon

Kimmeridjian (J_6) and Cenomanian (C_4) ages, and totaling about 900 m (18%) of
the total sequence thickness.

Even though, these two rock formations are composed of carbonate rocks (i.e.
in particular limestone and dolomitic limestone), yet they are characterized by
the secondary porosity and high permeability due to the abundant of fracture
systems of different scales. These fractures include fissure and multiple-set joints.

In addition, karstification is well developed among these two rock formations.
This includes rock galleries and subsurface conduits, as well as shafts and grottos which are almost filled with groundwater. It is often noticed that fracture
systems are usually accompanied with karstification (Shaban 2003).

2. Aquifer: Four rock formations were described as aquifers in the entire geologic
   sequence of Lebanon. They belong to rock formations of the Callovian (J_4),

Table 8.1 Aquiferous rock formations in Lebanon

Period	Rock formation[a]		Lithological characteristics	Aquiferous property	Thickness (m)
Quaternary	Holocene	Q	Alluviums and mixture of marine deposits and river terraces	Semi-aquifer	Variable
	Pleistocene		Alluvial, colluvial and thick soil deposits mixed with conglomerates		
Neogene	Pliocene	P	Marly limestone, and basalt with tuffs. joints are tremendous, notably in marly units	Aquitard	360
	Vindobanian	m_2-ncg	Conglomeratic limestone, and massive marly limestone and sometime with siltstone	Aquifer	320
	Burdigalian	m_1	Marly limestone and marl, sometimes exist as variable facies	Aquiclude	80
Paleogene	Lutetian	e_2	Limestone, chalky and marly limestone. Karistified and fractured	Aquifer	800
	Ypresian	e_1	Chalky limestone, marly limestone.	Aquiclude	370
Cretaceous	Senonian	C_6	Marl and marly limestone with joints	Aquiclude	400
	Turonian	C_5	Marly limestone and marl. Joints are predominant	Aquitard	200
	Cenomanian	C_4	Limestone, dolomitic limestone and marl. highly karistified and fractured	Excellent aquifer	700
	Albian	C_3	Marly limestone, marl and compacted shale	Aquiclude	200
	Upper Aptian	C_{2b}	Dolomite and dolomitic limestone. Highly jointed and fractured	Aquifer	50
	Lower Aptian	C_{2a}	Argillaceous sandstone, marl, limestone and clastic limestone	Aquitard	250
	Neocomian-Barremian	C_1	Quartzite sandstone, mixed with clayey and argillaceous materials. Highly jointed and sometimes variable and loose	Semi-aquifer	Variable
Jurassic	Portlandian	J_7	Oolitic limestone, with marly limestone	Aquitard	180
	Kimmeridjian	J_6	Dolomite, dolomitic limestone and limestone. Well karistified and fractured	Excellent aquifer	200
	Oxfordian	J_5	Mixed volcanics and tuff with marly limestone	Aquiclude	Variable
	Callovian	J_4	Dolomitic limestone and limestone. Karistified rocks	Aquifer	Undefined

[a]This represents the geologic epochs of the rock formations and their symbols (according to L. Dubertret 1955)

Upper Aptian (C_{2b}), Lutetian (e_2) and Vindobanian (m_2-ncg) ages, which are composed mainly of fractured and karistified carbonate that varying from dolomite and dolomitic limestone and marly and chalky limestone. They have a total thickness exceeding 1400 m (i.e. 28% of the total sequence).

These aquifers have regular hydrogeological properties that make them as groundwater-bearing rock masses, but there are some constraints occur and differentiate them from the excellent aquifers. This can be attributed to the following reasons:

– Callovian (J_4) aquifer has very limited areal extent in Lebanon and no complete exposure occurs.
– Upper Aptian (C_{2b}) aquifer has a limited thickness (i.e. less than 50 m), and it is often unable to retain groundwater because it comprises cliff exposures where groundwater easily outfalls from these exposures.
– Lutetian (e_2), described as Eocene aquifer and it is characterize by well bedded limestone with karstification and fractures, but the presence of chalk and marl in among the sequence of this aquifer reduces the aquiferous property.
– Vindobanian (m_2-ncg), usually described as Neogene aquifer. It is composed mainly from conglomeratic limes stone and marly limestone. Thus, conglomerates and marl slightly acting in reducing the aquiferous property.

3. Semi-aquifer: There are two semi-aquifers occur in the geological sequence. These are attributed to the rock formations of the Neocomian-Barremian (C_1) and Quaternary (Q). They do not have defined thickness, but the C_1 is usually ranges between 100 and 200 m, while Q has very diverse thickness that ranges from less than 1 meter to several ten of meters.

These rock formations were described as semi-aquifers and did not named as aquifers due to different reasons as follows:

– Neocomian-Barremian (C_1) is composed of interbedded sandstone and massive clay where thick sandstone sequence is rarely found to store considerable amount of groundwater. In addition, the mixture with clay reduces its porosity and permeability which are characterized the sandstone units. This why it has no define characteristics and cannot be described merely as aquifer.
– Quaternary (Q) has neither uniform thickness nor similar lithology. It is usually occur as mixture of different rock fragments and alluviums that mixed with soil. This results, in many instances, perched aquifers or aquifers with small- dimensions (i.e. thickness and extent), and then it is known by limited amounts of groundwater plus intermittent discharge.

4. Aquitard: There are six rock formations act in retarding groundwater flow. These are totaling about 1310 m (equivalent to about 26% of the entire sequence). They belong to: Pliocene (P), Vindobanian (m_2-ncg), Turonian (C_5), Lower-Aptian (C_{2a}), and Portlandian (C_7).

The majority of lithology for these formations implies the content of marl and marly limestone, and they are, in many instances, overlying with each other. These are: P and m_2-ncg, and thus forming a thick succession of groundwater retarding rock layers.

5. Aquiclude: Four rock formations comprise hydrologic barriers in the geologic succession in Lebanon. These are: Burdigalian (m_1), Senonian (C_6), Albian (C_3) and Oxfordian (J_5) where these formations have a total thickness does not exceed 700 m.

The clayey content, with minimal permeability, is the main factor in making these formations as constraints in groundwater flow whether in the vertical or horizontal direction.

8.2.2 Hydraulic Properties

Hydraulic properties describe the flow behavior within the rock formation or unit itself. It is also the collective hydraulic characteristics of rock or soil. According to Brassington (1990), the characteristics which control groundwater mechanism (including flow and storage) are usually referred to as the hydraulic properties which can be measured in the field or laboratory, but can also be assessed in general terms by consideration of the overall aquifer geology.

Thus, the description of hydraulic property differs from the aquiferous property in that the latter focuses on water interface between different rock formations while the hydraulic one often concentrates on water movement within the same rock unit, and it can be to very limited distance. Whereas the hydrologic and aquiferous properties have almost similar effect when describing water content and flow between different rock formations.

There are many hydraulic properties applied to investigate the characteristics of soil and rock. Most of these properties were not precisely calculated in Lebanon, and conflicting estimates are often found. In particular, the following ones are the most significant properties:

1. Porosity (Ø): It is the proportion of the volume of voids and openings in a rock to the volume of that rock, and it is usually expressed as percentage. While the "effective porosity" is the total porosity excluding the isolated pores. Hence, porosity is controlled by grain size and shape which are created during the sedimentation of these grains, like in the case of the sandstone and clay with are occupied in the geologic succession in Lebanon.

   Secondary porosity is also another aspect of porosity, but it referred to the openings in rocks which are resulted from rock deformations (e.g. fractures, etc.). This is common in Lebanon where most of the carbonate rocks are highly fractured and then characterized by the secondary porosity.

2. Permeability (ρ): It expresses the capacity of a *rock* to transmit fluid, and it is a function of resistance to the flow of a fluid through a rock. Rock structures play a role in creating openings among which fluid moves. It is almost accompanied with the secondary porosity, such as the case of the porous and permeable carbonate rocks in Lebanon.

   Permeability usually follows different descriptive classifications; therefore, highly permeable, moderately, etc. are examples for describing permeability.

3. Moisture content (ŋ): This is the quantity of water contained in soil or rock specimen or layer. It is calculated as a percentage between dry and wet weight samples. The moisture content is mainly a function of porosity but not the permeability. Thus, ŋ can be calculated as the following formula:

$$\text{ŋ} = \left(W_w - W_d \right) / W_w \times 100$$

Where W_w is the wet rock specimen and W_d is the dry rock specimen.

4. Hydraulic conductivity (k): This property represents the ability of a fluid to flow through the pores or fractured rocks. In more certain definition, it is the volume of water that will flow through a unit cross-sectional area of aquifer in unit time. Hence, it can be expressed by the following equation:

$$Q = A \times k \times h / l$$

Where Q is the water volume flowing in unit time through the sectional-area A, and k is the hydraulic conductivity (m/sec) and h/l is the hydraulic gradient.

5. Specific yield (S_y): This is the ratio of the volume of water that a saturated rock will yield by gravity to the total volume of the rock. Thus, it can be expressed by the following equation:

$$S_y = V_d / V_t$$

Where V_d is the volume of water that drains from the total volume of rock V_t.

6. Transmissivity (T): It is represented by the rate of flow under unit hydraulic gradient through a cross-section of unit width over the whole saturated thickness of the aquifer (Kruseman and de Ridder 2000). It is; therefore, calculated in m²/sec, by the following formula:

$$T = k \times d$$

Where k is the hydraulic conductivity and the saturated thickness d (m).

Table 8.2 shows general estimations about the hydraulic properties for the aquiferous rock formations in Lebanon. These estimations have been adopted either from field measures/or from different sources (e.g. Tabet 1978; Canaan 1992; Shaban 2003; Khadra 2005; Brassington 1990; Kruseman and de Ridder 2000). While, Fig. 8.2 shows streamlined description for these hydraulic properties.

8.3   Estimated Aquifers' Capacity

Since there are contradictory estimates in the hydraulic properties of rock masses in Lebanon, thus the volume of groundwater is also undefined and many diverse values were put by different researchers. In this respect, the National Water Sector Strategy

Table 8.2 Hydraulic properties of aquiferous rock formations in Lebanon

Rock formation[a]		Hydraulic properties					
		Ø	P[a]	η	k	S_y	T
		%	–	%	m/sec	%	m^2/sec
Quaternary	Q	8–18	L-M	5–15	–	–	–
Neogene	m_2-ncg	6–10	M	8–10	–	12–18	–
Lutetian	e_2	4–6	M-H	7–9	10^{-5}–10^b	14–24	1.2×10^{-6} – $1. \times 10^{-6}$
Cenomanian	C_4	8–12	H-VH	6–8	10^{-5}–10^b	16–26	1.8×10^{-6} – 2.4×10^{-6}
Upper Aptian	C_{2b}	14–16	H-VH	4–8	10^{-5}–10^b	26–38	2.3×10^{-6} – 3.2×10^{-6}
Neocomian-Barremian	C_1	11–15	M-H	12–16	10^{-4}–10^b	25–30	4.2×10^{-5} – 7.1×10^{-5}
Kimmeridjian	J_6	8–12	H-VH	6–8	10^{-5}–10^b	14–26	1.5×10^{-6} – 2.2×10^{-6}
Callovian	J_4	5–8	M-H	7–9	10^{-5}–10^b	10–20	2.5×10^{-6} – 5.9×10^{-6}

[a]*L* Low, *M* Moderate, *H* High, *VH* Very high
[b]Values at the higher end of the range occur where there is secondary porosity or permeability

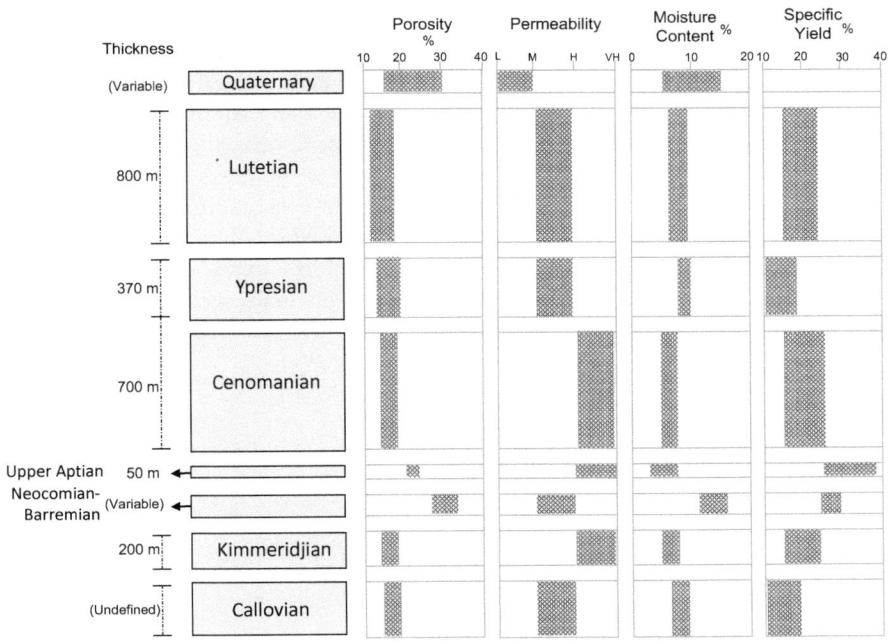

Fig. 8.2 Estimated hydraulic properties of aquiferous rock formations in Lebanon as deduced from several measures and different sources

estimated the total water volume in Lebanon at 2.7 billion m³ where 2.2 and 0.5 billion m³ are allocated to surface water and groundwater; respectively (MoEW 2010). Besides, it was reported that the total groundwater abstraction by wells is estimated at 0.70 billion m³, it is considered that under normal climatic conditions there is a yearly deficit of 0.2 billion m³ in groundwater (IWMI and USAID 2017), and this must not quite accurate and it gives a measure of the degree of uncertainty about the resource itself. The estimation illustrated by FAO (2008), shows that groundwater recharge in Lebanon is estimated at 3.2 billion m³. While, MoEW and UNDP (2014) has come up with much higher values for groundwater volume where it was estimated that there is 3.57 billion m³ in a dry year and 6.1 billion m³ in a wet year (averaging 4.84 billion m³).

There are eight rock formations in Lebanon that are considered as aquiferous rock units, but with different hydrgeologic characteristics and diverse potentiality for groundwater storage (Tables 8.1 and 8.2). Six of these aquiferous rock units are mainly composed of fractured and karistified carbonate rocks which belong to the Jurassic, Middle Cretaceous and Eocene ages, while one is clastic unit (Neocomian-Barremian – C1) and the other is of Quaternary deposits (Q).

The total thickness of these rock units exceeds 3000 m which is equivalent to 60% of the entire geologic sequence of Lebanon. While, they occupy between about 8332 km² (80%) of the Lebanese territory. Nevertheless, these two variables (thickness and area) will be less when calculating the aquifers capacity, and this can be attributed to the effective thickness of any aquifer according to the localities where it is situated. In other words, some aquiferous rock formations are found in specific areas, but this areas (sometimes) include minimal thickness which is not supposed to hold groundwater; therefore, the whole identified area will not be considered.

The areal extent (i.e. exposed and wrapped) of the major aquiferous rock formations are listed in Table 8.3. These were calculated following the preliminary approaches done by Shaban (2003, 2012).

Adopting this concept, therefore the total area for each aquiferous rock formation was estimated by measuring the area of the exposed lithologies of these formations, plus the estimated areal extent of the covered lithologies for the same rock formations. This extent was assessed depending on many geological elements including mainly the: (1) lithological characteristics, (2) interbedding and contacts

Table 8.3 Areas of the aquifers rock formations in Lebanon

Aquiferous unit		Mount Lebanon	Anti-Lebanon	Bekaa plain
Quaternary		Varies (up to 1466 km²)		
Neogene		683 km²	–	265 km²
Eocene	Lutetian	684 km²	–	102 km²
Cretaceous	Cenomanian	2615 km²	450 km²	1060 km²
	Upper Aptian	181 km²		3 km²
	Neocomian-Barremian	436 km²	–	–
Jurassic	Kimmeridjian	1333 km²	159 km²	15 km²
	Callovian	315 km²	31 km²	–

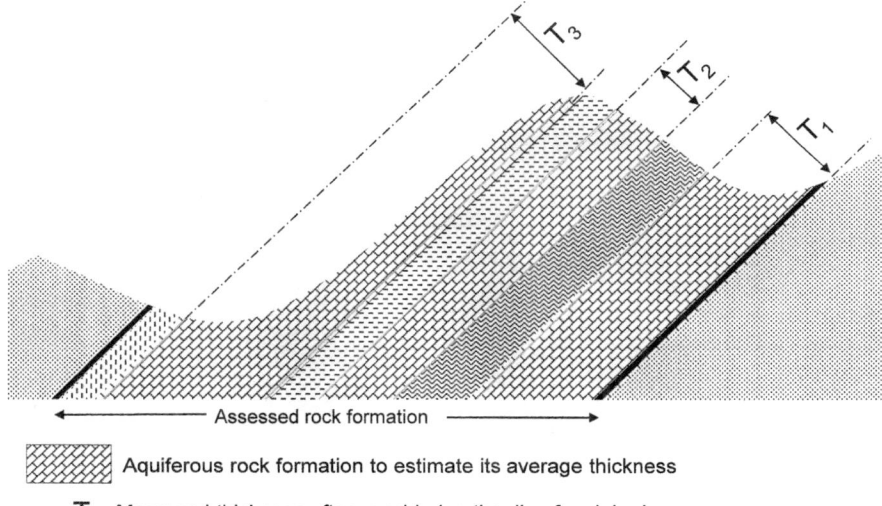

Assessed rock formation

Aquiferous rock formation to estimate its average thickness

T   Measured thickness after considering the dip of rock beds

Fig. 8.3 Example for calculating the average thickness

of rock formation, (3) geologic structure (in particular, domes, folds and fault alignments) and (4) location of different rock exposures.

For the thickness mentioned in Table 8.1, it reveals the cumulative thickness of each rock formations. However, the average thickness is different from that of the cumulative one. Therefore, the average thickness was calculated taking into accounts the thickness variation with respect to the areal extent, dip of bedding plans, as well as the non-aquiferous units were excluded. Figure 8.3 shows an example on how the average thickness was calculated.

Therefore, the estimation of the potential groundwater-bearing capacity (A_c) in each of the aquiferous rock formations can be applied when the three variables for the area (A), average thickness (T) and porosity (Ø) become ready according to the following simplified formula:

$$A_c = A \times T \times \phi$$

The three variables are illustrated in Tables 8.1, 8.2, and 8.3. Therefore the estimated capacity in the aquiferous rock formations can be calculated and the resulted are shown in Table 8.4. It is clear that the majority of groundwater capacity is for the Cenomanian rock formation with approximately 54,855 million m^3. Then it followed by Kimmeridjian and Lutetian rock formations with 10,780 and 6120 million m^3; respectively.

The resulted estimation of the potential groundwater-bearing capacity, which is 88.9 billion m^3, indicates the capability of the aquiferous rock formations in Lebanon

Table 8.4 Estimated volume of water in the aquiferous rock formations in Lebanon

Aquiferous rock formations		Area[a] (km^2)	Thickness[b] (m)	Porosity $(\%)$	Estimated aquifers' capacity (Mm^3)
Quaternary		775[c]	5 (variable)	8–18	456
Neogene		948	60	6–10	4550
Eocene	Lutetian	786	150	4–6	5895
Cretaceous	Cenomanian	4125	150	8–12	61,875
	Upper Aptian	184	10	14–16	276
	Neocomian-Barremian	436	40 (variable)	11–15	2267
Jurassic	Kimmeridjian	1507	80	8–12	12,056
	Callovian	346	60 (undefined)	5–8	1557
Total					**88,932**

[a]Area of the exposed and covered rock masses (According to Table 2.3)
[b]Average effective thickness (potential to store groundwater) of rock formations
[c]Quaternary deposits with considerable thickness

to store groundwater. However, only 4.84 billion m^3 is adopted as a reliable volume of groundwater in Lebanon. This means that groundwater occupies only 5.5% of the capacity of the Lebanese aquifers.

8.4 Groundwater Recharge

The recharge of groundwater from terrain surface into substratum is always considered as a major factor in many studies. This is because the "recharge" as a hydrologic process controls the mechanism of water flow which is a significant factor used while exploring groundwater, positioning suitable sites for constructions and the related environmental features (e.g. landfills). Also, recharge is usually accounted while calculating the water budget. In this respect, UNDP (1970) estimated that the recharge of water into the underlying rock exceeds 40% of the precipitated water. While, Abbud and Aker (1986) put several estimates for the recharge though rock formations in Lebanon starting from 0% to 41%. This reflects the importance of determining more precise estimates for this influencing hydrologic process.

Yet, there is a contradictory in the discrimination between infiltration, percolation and recharge. Where the first is often used while assessing the water flow through soil and the second is used to focus on the feasibility of water transit into rock masses for localized sites. While, "recharge" describes water flow rate from surface into subsurface as viewed from a geospatial distribution and usually applied for large-scale areas.

For groundwater exploration, it is necessary first to identify the surface zones wherefrom water can easily recharged downward into the underlying rock masses. It would be the first clue to for tracing the water journey and to end up with the most probable location where groundwater may store (Shaban 2003, 2010).

Nevertheless, some researchers depend in their studies only on faults to identifying the recharge zones (Gustafsson 1994; Teeuw 1995), others involve faults and drainages (Ahmed et al. 1984; Edet et al. 1998; Robinson et al. 1999); besides many include faults, drainages and lithology as the principal elements to identify re-charge potential zones (Bilal and Ammar 2002; Kumar et al. 2007). There are also researchers who used several features to identify the recharge potential including, in addition to the mentioned ones above, land cover/use, slope, etc. (Ganapuram et al. 2008; Al Saud 2010).

The determination of groundwater recharge zones in any area is controlled by several factors where these factors differ from one region to another and sometimes they have different weights of influence between diverse regions.

8.4.1 Factors Controlling Recharge Rate

According to the geographic location of Lebanon, five controlling factors are considered for groundwater recharge assessment (Shaban et al. 2005). These are: fracture systems, drainage systems, lithological characteristics, karstification and land cover/use.

1. Fracture systems:

Rock deformations, and certainly the fracture systems, are utmost significant controlling factors in increasing the recharge rate from surface to sub-surface stratums. These systems include fissures and joints and even the contacts between bedding plans. However, for the large-scale assessment and mapping of the recharge potentiality, usually faults are determined, because other fracture systems are often occurred within local dimensions and limited localities. In the regard, faults are also accompanied with fracture systems of different aspects and dimensions (Shaban 2003).

The identification of faults, for the large-scale recharge assessment, is done by using remote sensing techniques in combination with the available geologic maps (Fig. 8.4). Therefore, the identified faults on satellite images are known as "lineament" (Khawlie and Shaban 1998; Shaban and Khawlie 2006).

The elaborated research, by the author for mapping the recharge potential zones for Lebanon, lineaments were recognized from satellite images of Landsat 7 ETM (30 m spatial resolution) and Aster (15 m spatial resolution). Thus, the ERDAS-Imagine software was used for the processing of these images, notably that this software has the digital advantages of detecting edges of the observed objects, and it enables the thermal recognition of linear features on the processed satellite images.

The concept behind calculating the density of lineaments, i.e. faults (D_l) implies the following formula:

$$D_l = N_l / A$$

Where N_l is the number of lineaments within a specific area (A).

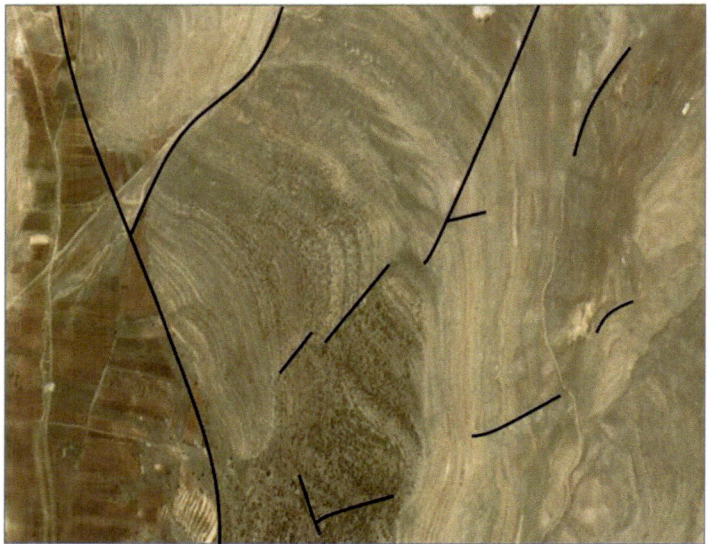

Fig. 8.4 Example showing the identified faults (black color) on Aster satellite image

A contour map, showing the density of the recognized lineaments, was achieved using "Sliding Windows" approach (Shaban et al. 2005). Hence, five classes were resulted to describe the lineaments' density with respect to recharge rate.

2. Drainage systems:

There are several morphometric characteristics used to investigate the behavior of drainage system towards the hydrological processes, such as: drainage density, drainage frequency and the density of streams intersection, length, bifurcation ratio, drainage pattern, etc. However, drainage density is still a principal element that is often analyzed to investigate the recharge rate. Thus, the denser the drainage network, the less recharge rate and vice versa.

In this respect, the tracing the drainages network is the primary phase for further drainage density elaboration as done by the author's studies (Shaban et al. 2005; Shaban and Khawlie 2006). Thus, the drainages (major and minor streams) were extracted from the Global Digital Elevation Model (GDEM) as generated by the space-borne satellite ASTER with 30 m resolution. This was supported by the available topographic maps (1:50000 scale). Hence, Arc-GIS software, through it Arc-Map, Arc-Catalogue and Arc-Toolbox applications, was used to generated the detailed drainage network for Lebanon.

When the drainage network map has been produced; therefore, similar concept used in lineament density was used for drainage density (D_d), but here the total length of streams (N_s) was calculated in a defined area (A) as follows:

$$D_d = N_s / A$$

Therefore, five classes of drainage density were generated and then mapped in a contour map form for further data manipulation.

3. Lithology:

Other than the deformations that occur in rocks, no doubt, the lithological characteristics have a fundamental role in water transit from surface to sub-surface, as well as within the rock mass itself or between different rock beds. This implies the degree of primary porosity and permeability included in the rock masses.

Lithological characteristics represent the interior behavior of water regime into rock masses, there many studies accounted the lithology and faults as the major two clues for groundwater exploration (Savane et al. 1996; Sener et al. 2005).

According to the existing rock lithologies in Lebanon, they can be characterized by diverse water-flow responding behavior as it was listed in Table 8.1. Therefore, the dominant carbonate rocks are known by well responding to water flow (i.e. good recharge) as a result of secondary porosity and permeability, such as for the Cenomanian limestone. However, the rate of recharge in these lithologies is reduced once they contain marly and clayey facies as the case of the Turonian marly limestone for example. The other lithologies which contain considerable amounts of marl and argillaceous materials have poor responding to water flow (e.g. Senonian marl). While, the clastic rocks, in particular the Neocomian-Barremian sandstone they have good recharge property unless the clay masses intruded among these rocks.

Based on the above, the exposed rock formations in Lebanon were sorted according to their responding to water flow. Hence, five classes were elaborated and mapped for this purpose.

4. Karstification:

As it was discussed in Chap. 2, more than 85% of the carbonate rocks in Lebanon are karistified (i.e. subjected to dissolution of carbonate rocks) with a variety of aspects including surface and sub-surface rocks and even the contact between them. Thus, contact-surface karstification is a major clue for water transit from surface to sub-surface. It is represented mainly by the sinkholes and many other aspects of water-collect rock pockets. Therefore, these features were considered by the author as potential localities for water entrance into the underlying rock masses (Shaban 2003; Shaban and Darwich 2011).

It was also noted that these surface features, which represent openings to the sub-stratum, are usually connected with sub-surface structures, such as fault alignments and intensive fracture system (example in Fig. 8.5).

5. Land cover/use:

Other than the exposed rock masses, the features forming Earth's surface are also significant in controlling water flow whether on terrain surface or downward to the underlying stratum. These features can be of natural origin or made by human where they are changeable with time and have different responding to water recharge.

There are several classifications for the LCU. These were discussed in details in Chap. 2 (Sect. 2.5). Thus, the majority of LCU implies generally: bare lands,

Fig. 8.5 Set of sinkholes along wrapped fault alignment in Mount-Lebanon

agricultural areas, forests, urban areas and water bodies. Under each of these components, there are many sub-divisions and the number of classes is always dependent on the purpose of study. Therefore, the used map for the recharge property is
the same as that was elaborated in Chap. 2, where 8 classes were adopted.

8.4.2 Data Manipulation

There are five maps prepared digitally for the factors influencing the identification
of recharge potential zones in Lebanon as discussed in Sect. 8.4.1. Each of these
maps represents one "GIS layer" for one factors. Therefore, the integration of all
maps into the GIS system will result the optimal (& final) map for the recharge
potential zones. This digital integration (superposition) can be applied by using
ESRI's Arc View software.

It must be made clear that not all these factors influence at the same level of impact;
therefore, a number of weights were given to each of them. Hence, the determination of
weights was followed several applied approaches (Khawlie 1986; Shaban et al. 2001).
However, the most creditable weighting approach for the factors influencing the recharge
process was adopted by Shaban et al. 2005. This weighting was as follows:

- Faults (Lineaments) = 26%
- Drainage = 16%
- Lithology = 26%
- Karstic domains = 21%
- Land cover/land use = 11%

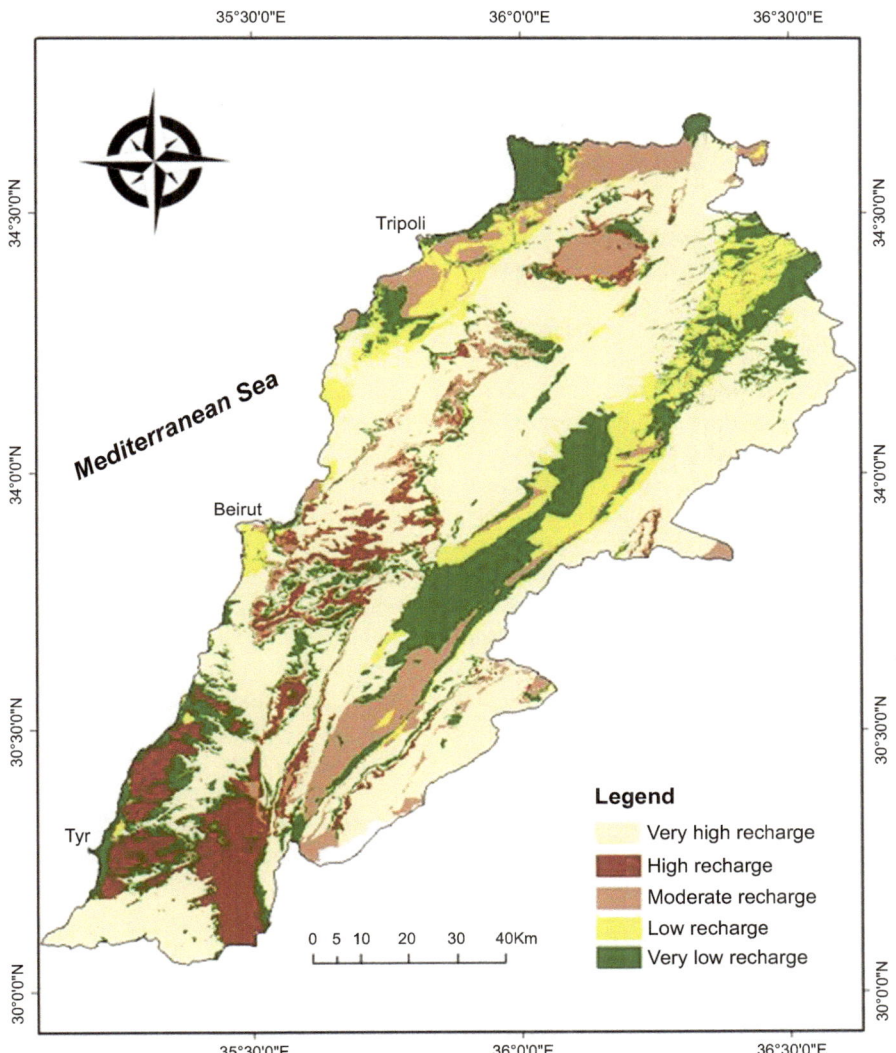

Fig. 8.6 Recharge potential maps for Lebanon

Based on the above weights, and the digital integration of different GIS layers for the factors influencing the recharge rate; however, the recharge potential map was produced for the entire Lebanon (Fig. 8.6.).

The produced map reveals the major zones for the recharge potential with five descriptive levels, ranging from very high, high, to very low. Therefore, it was found that 51% of the Lebanese territory is characterized by very high recharge potential (Fig. 8.6).

Also, it was calculated that the areas of recharge zones are: 12%, 11%, 10% and 16% for the high, moderate, low and very low rates; respectively. This pointed out that the Lebanese territory is characterized by 63% (almost 2/3) of considerable recharge rates, which integrally contributes in recharging groundwater reservoirs.

8.5   Groundwater and Faults

In the view of the existing water shortage, it becomes challenging to explore groundwater resources. Thus, there are several methods applied to recognize the potential localities for groundwater. However, positive results sometime occur and groundwater can be found with considerable amounts, but this is not always the case and some dug wells were failed because no groundwater was found or the explored groundwater was very little and with intermittent discharge. This motivated hydrogeologists to adopt credible approaches to while applying ground surveys to recognize groundwater potential zones.

In this respect, fracture systems are always given the most concern. This is based on the concept that groundwater storage needs porous rock masses, and then fractures enhance the secondary porosity in these rock, exactly such as the case in the carbonate rocks of Lebanon. Therefore, the analysis of fractures is the primary clue used. However, not all these fractures make it feasible for groundwater storage, and there are many elements must be considered in this respect. Hence, not all aspects of fractures play the positive role in groundwater storage. In fact, sometimes fractures are good elements for groundwater recharge as mentioned in Sect. 8.4., they always contribute in the increasing of porosity and permeability; besides they can transit water from one rock mass to another and reduces the storage rate.

The author noticed a good clue while localizing potential site to drill for groundwater. That is the existence of faults which became the major element of groundwater exploration. This is well pronounced in Lebanon, notably that faults create elongated zones of high porosity and permeability even in compacted rock masses (Shaban 2003). For this reason, the author and other applied an empirical investigation to deduce the relationship between faults and groundwater (Shaban et al. 2007). The applied investigation was on four pilot areas in Lebanon that representing different physical setting including mainly diverse geology.

8.5.1   Elements of Investigation

1. Faults:

There was a discussion on faults in Sect. 8.4; however, for deducing the relationship of faults for groundwater can be viewed from describing faults dimensions and aspects. Therefore, faults can be of different scales where they can be few meter up to thousands of kilometers. Also their displacement is either vertical, horizontal or diagonal, and thus norm, thrust and strike-slip faults occur.

The recognition of faults, as discussed in Sect. 8.4, was dependent on the available geological maps (1:50000), satellite image analysis and the field verification (Shaban et al. 2005). For the satellite images processing, more in-depth analysis was applied for the purpose of identifying the relationship between faults and groundwater. Thus, lineaments (geologic origin features) were the recognized.

Landsat 7 ETM and Aster images were processed using ERDAS-Imagine, and many digital advantage were used to verify the optical and thermal observations on these images. This includes: edge detection, directional filtering, contrasting and sharpness. In addition, band combination war performed and the ordering of bands 2, 5 and 3 on Landsat images was found to be the best in recognizing linear features. Moreover, the thermal interpretation from thermal band, i.e. band-6 (120 m resolution), was also undertaken. Therefore, lineaments in the four pilot areas were extracted (Shaban et al. 2007).

2. Water wells:

Groundwater elements were represented by the investigation of drilled water wells where 90 water wells were surveyed in four pilot area. Thus, hydrogeological data on these wells were collected with a special focus on, depth of water table, discharge and the rock formation where they belong to. However, for the scope of investigation, the discharge rates (water productivity in well) were also considered.

Therefore, wells' productivity was described as very high, high, moderate, low and very low for the discharges of >20, 19–15, 14–10, 9–5 and <5 l/sec; respectively.

8.5.2 Data Analysis

There are many hydrogeological clues used to recognize groundwater potential zones. One of them, for example, based on the understanding which was adopted by the author and it implies three aspects of fault/lithology interrelationship to groundwater potentiality as shown in Table 8.5. Hence, the location of two lithologies with different perviousness characteristics along a fault alignment is often potential to store groundwater even though this may create flow of groundwater with considerable amounts.

The above concept can be used while selecting suitable sites for groundwater storage, but it cannot be adopted for precise identification of the relationship between faults and groundwater productivity. Therefore, faults must be analyzed from the dimensional point of view. This implies verifying the relationship between the locations of water wells and faults' (i.e. lineaments). For this purpose, the author and others (Shaban et al. 2007) primarily used two imperial formula, as follows:

1. Location of water wells with respect to faults' density (F_d), i.e. sum of faults length over area.
2. Location of water wells with respect to faults' frequency (F_f), i.e. sum of faults number over area.

Table 8.5 Groundwater storage and flow with respect to lithologies along faults

Fault location *(between)*	Groundwater storage[a] *(along fault)*	Groundwater flow[b] *(along fault)*
Pervious/pervious lithologies[c]	Moderate	Low
Impervious/ impervious lithologies	Negligible	Moderate – high
Pervious/impervious lithologies	Effective	Moderate – high

[a]Storage in considerable volume
[b]Flow can be lateral and horizontal
[c]Pervious or impervious lithologies can be for the same rock formation

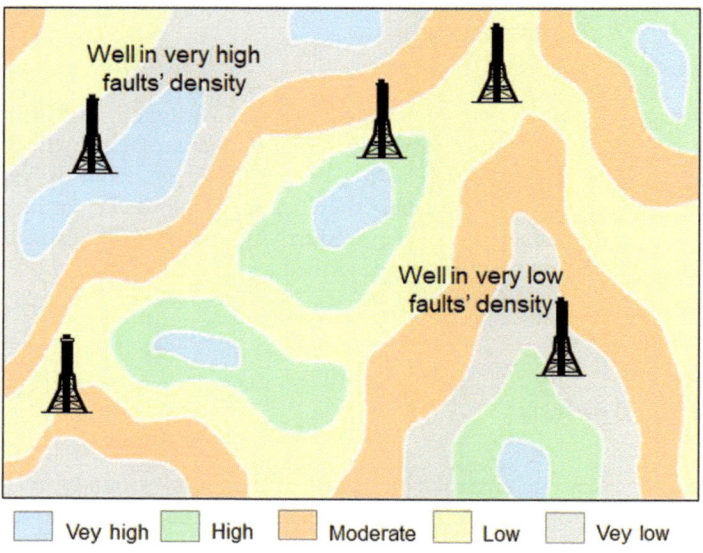

Fig. 8.7 Example showing the location of wells on faults density map

Thus, the location of the 90 surveyed wells was plotted on maps representing the above two formula (example: Fig. 8.7). While, Table 8.6 shows the illustrated data belong to these formula. Therefore, the resultant showed that the coincidence is 20% and 27% for faults density and faults frequency; respectively. This does not make proper evidence of any relationship between water productivity and wells and according to the two investigated formula.

Another empirical approach was used to investigate the relationship between faults and groundwater productivity. This focuses on verifying the proximity of fault(s) with respect to the productive wells. In other words, the distance factor between faults and the successful water wells. For this purpose, the locations of the 90 water wells were projected on faults map using the Arc-Info (in Arc-GIS). The

Table 8.6 Groundwater productivity and faults dimensional characteristics

#	RF[a]	GWP[b] (l/sec)		Faults characteristics		
				(F_f)	(F_d)	(F_p)[c]
1	e_2	15	HP[d]	H	M	210
2	e_2	21	VHP	H	M	75
3	e_2	4	VLP	H	H	1350
4	C_{2a}	15	HP	L	L	250
5	C_4	20	VHP	M	M	50
6	C_{2a}	4	VLP	L	M	775
7	J_6	10	MP	M	H	650
8	J_6	23	VHP	H	M	150
9	J_6	21	VHP	VH	H	105
10	J_6	5	LP	M	H	340
11	J_6	4	VLP	M	M	850
12	J_6	20	VHP	M	H	125
13	J_6	16	HP	H	M	305
14	C_{2b}	8	LP	L	L	840
15	C_4	12	MP	L	VL	1200
16	C_4	20	VHP	H	H	50
17	C_4	11	MP	H	M	100
18	e_2	0	VLP	H	VH	1500
19	C_4	16	HP	M	H	20
20	C_4	17	HP	H	M	500
21	C_4	20	VHP	L	L	50
22	C_4	18	HP	M	L	70
23	C_4	22	VHP	H	H	110
24	C_4	11	MP	L	L	220
25	C_4	17	HP	M	M	10
26	e_2	2	VLP	M	H	2500
27	C_4	0	VLP	VL	VL	1200
28	C_4	12	MP	M	H	650
29	C_4	16	HP	M	M	20
30	C_4	9	LP	M	M	1425
31	C_4	7	LP	L	M	420
32	C_4	22	VHP	M	M	10
33	e_2	25	VHP	H	H	15
34	C_4	21	VHP	M	H	5
35	C_4	20	VHP	H	M	200
36	C_4	0	VLP	H	M	1600
37	C_4	22	VHP	M	H	15
38	C_4	8	LP	M	H	370
39	C_4	10	MP	L	H	465
40	C_4	17	HP	L	M	270
41	C_4	21	VHP	H	L	130

(continued)

Table 8.6 (continued)

#	RF[a]	GWP[b] (l/sec)		Faults characteristics		
				(F_r)	(F_d)	(F_p)[c]
42	C_4	10	MP	M	VL	365
43	C_4	12	MP	M	L	780
44	C_4	9	LP	M	L	955
45	C_4	15	HP	VH	H	245
46	C_4	11	MP	M	H	660
47	C_4	10	MP	L	M	735
48	C_4	7	LP	L	M	1835
49	C_4	5	VLP	L	L	1645
50	C_4	6	LP	L	L	1210
51	C_4	6	LP	M	VL	1405
52	C_4	6	LP	L	H	1350
53	C_4	12	MP	L	M	620
54	C_4	10	MP	VL	L	410
55	C_4	6	LP	VL	L	525
56	C_4	7	LP	H	M	565
57	C_4	6	LP	M	M	505
58	C_4	10	MP	M	H	515
59	Q	11	MP	H	VH	550
60	C_4	10	MP	VH	VH	405
61	Q	6	LP	L	M	175
62	Q	5	VLP	H	L	1815
63	Q	7	LP	H	M	2075
64	C_4	11	MP	VH	H	215
65	C_4	15	HP	H	M	135
66	C_4	16	HP	M	L	140
67	C_4	15	HP	L	VL	50
68	C_4	20	VHP	VH	H	80
69	C_4	35	VHP	M	M	20
70	C_4	35	VHP	L	H	10
71	C_4	22	VHP	L	M	55
72	C_4	16	HP	VL	L	95
73	C_4	10	MP	M	M	305
74	C_4	16	HP	M	H	225
75	C_4	17	HP	H	M	125
76	C_4	11	MP	H	H	195
77	C_4	12	MP	H	L	385
78	C_{2b}	12	MP	M	L	360
79	C_{2b}	12	MP	L	L	275
80	J_6	15	HP	L	VL	200
81	Q	0.5	VLP	L	VL	1655
82	Q	0.5	VLP	VL	L	1675

(continued)

Table 8.6 (continued)

#	RF[a]	GWP[b] (*l/sec*)		Faults characteristics		
				(F$_f$)	(F$_d$)	(F$_p$)[c]
83	Q	0.5	VLP	VL	M	990
84	Q	0.5	VLP	L	M	1098
85	Q	1	VLP	M	M	1955
86	Q	0.5	VLP	M	H	1123
87	Q	0.5	VLP	H	H	1345
88	Q	0	VLP	M	L	2790
89	Q	0.25	VLP	L	VL	1825
90	Q	0.25	VLP	M	VL	1635

[a]*RF* Rock formation
[b]*GWP* Groundwater productivity
[c]Distance between a fault and a water well
[d]*H* high, *V* very, *M* moderate, *L* low

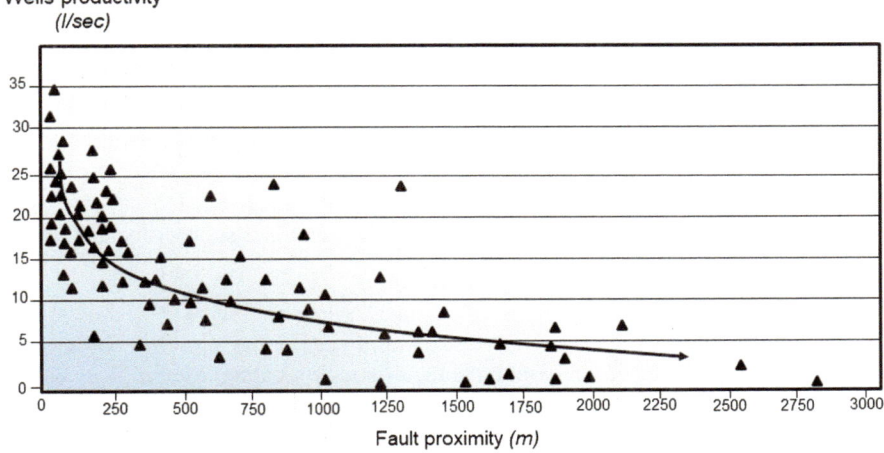

Fig. 8.8 Relationship between faults proximity and groundwater productivity in wells

distance between both, i.e. proximity (F$_p$) was calculated regardless of the side where they are located with respect to each other. In particular, this was applied to relatively faults with considerable dimension and not to local faults with limited geological effect. Hence, resulted data are put in Table 8.6.

Data on faults proximity were plotted graphically to deduce the relationship between faults distance and water productivity in wells (Fig. 8.8). Consequently, it was found that dug wells located in the proximity of less than 350 m from faults are characterized with water productivity exceeding 15 *l*/s. While, wells located at a distance between 350 and 650 m from a fault usually show productivity of around 10 *l*/s. Therefore, more than 97% of the wells which are located in the proximity or less than 650 m from a faults encompass water productivity larger than 10 l/sec. This brings the conclusion that the more proximity of wells to a fault structure, the higher productivity occurs and vice versa.

8.6   Groundwater Level and Discharge

Among the approximately 5000 m of the total thickness of the Lebanon's stratigraphic succession, there are different groundwater levels (or water tables) and discharge rates. It is; therefore, essential to determine groundwater level and discharge in any area, because they are usually used as a function of the dimension of groundwater reservoirs. In addition, depth and discharge rate of groundwater are always viewed from the financial point of view, thus land owners are highly concerned in identifying the depth to groundwater before drilling wells; in addition, water discharge is also of significance up on which they can take their decision. Also, the water table is often indicative to characterize the contamination in aquifers.

Groundwater level and discharge are dependent on several factors as follows:

1. Positioning of aquiferous rock units within the same rock formations (e.g. upper, middle, etc.),
2. Existing geologic structures (e.g. faults, anticline, etc.),
3. Permeability and porosity of the existing rock masses and the interbedding aspects of these rocks,
4. Altitude and the accompanied geomorphological features,
5. Water feeding rate including precipitation and other water sources located in the proximity,
6. Pumping rate regime of groundwater abstraction.

The aquiferous rock formations were determined in the previous discussion (Sect. 8.2), and more specifically their characteristics were diagnosed in Sect. 8.2.1. However, there are several units included in each aquiferous rock formation. These units are sometimes characterized by different hydrologic properties than the entire rock formation. In other words, the description of each rock formation is usually applied for the general lithological aspects and not specifying the detailed units included among it.

8.6.1   Groundwater Level

The following is a diagnoses for the aquiferous rock formations in Lebanon, based on Table 8.1 and Fig. 8.1:

1. Quaternary Aquifer:

The lithology of the Quaternary deposits, as a mixture of soil and detrital rocks with disintegrated particles makes it undefined in terms of the aquiferous property.

Therefore, the thickness of the Quaternary deposits in Lebanon varies between few meters up to 40 m where the maximum thickness is reported in the Bekaa Plain (averaging 15–20 m). Thus, the thicker Quaternary deposits are found, in addition to the Bekaa plain, in the coastal plains.

The majority of the lithological characteristic of the Quaternary deposit includes mainly alluvial deposits mixed with clayey soil. Thus, groundwater is stored mainly within the alluvial deposits and retained by the clayey sediments Therefore, several groundwater levels exist among these deposits as shown in Table 8.7. These levels are only within the Quaternary deposits, while another (& deeper) water levels can be found in the same localities, but they belongs to another aquiferous formation.

2. Neogene Aquifer:

The Neogene aquifer, mainly of the Miocene age (m_2-ncg) is widespread in the Bekaa Plain and in Al-Koura Region, north Lebanon. It is composed mainly of locally interbedded conglomeratic limestone and massive marly limestone, which are dominant with joints and local fissures. The vertical interbedding between these two rock units is frequently occurred (Fig. 8.9). Therefore, water is accumulated at different levels in the conglomeratic limestone where the marl and marly limestone act as impermeable layers. In other words, the Neogene has several groundwater depths even in the same locality. This is well pronounced in the dug wells located in the Bekaa Plain.

According to the hydro-stratigraphic sequence of the Neogene Aquifer, the main water table levels is almost at 120–140 m in the region of Bekaa (Shaban 2017). While a number of perched water tables occur at different levels (Fig. 8.9).

3. Lutetian Aquifer:

This is the Eocene Aquifer which is composed largely of we bedded, moderately thick to thin beds of limestone. It is highly fractured and jointed, as well as karistified and almost including chert nodules. This limestone is interbedded with marly and chalky limestone at different depths. Therefore, groundwater is located mainly

Table 8.7 Groundwater level in the Quaternary deposits

Geographic distribution	Major lithology	Estimated groundwater level
Bekaa Plain	Relatively thick sequences of loamy soil and alluviums	2–15 m
Akkar Plain	Colluvial deposits, silt and sand mixed with alluviums	5–25 m
El-Menieh	Sand dunes mixed with alluvial deposits	5–15 m
Chekka	Alluvial deposits	3–10 m
Damour Plain	Loamy soil and alluviums	5–30 m
Al-Aakbieh	Mixed loamy soil mixed with white Rendzina	15–40 m
Adloun-Ain Abou Abdallah		10–25 m
Rachidieh-Ras Al-Ain	Silt and alluvial deposits	10–15 m

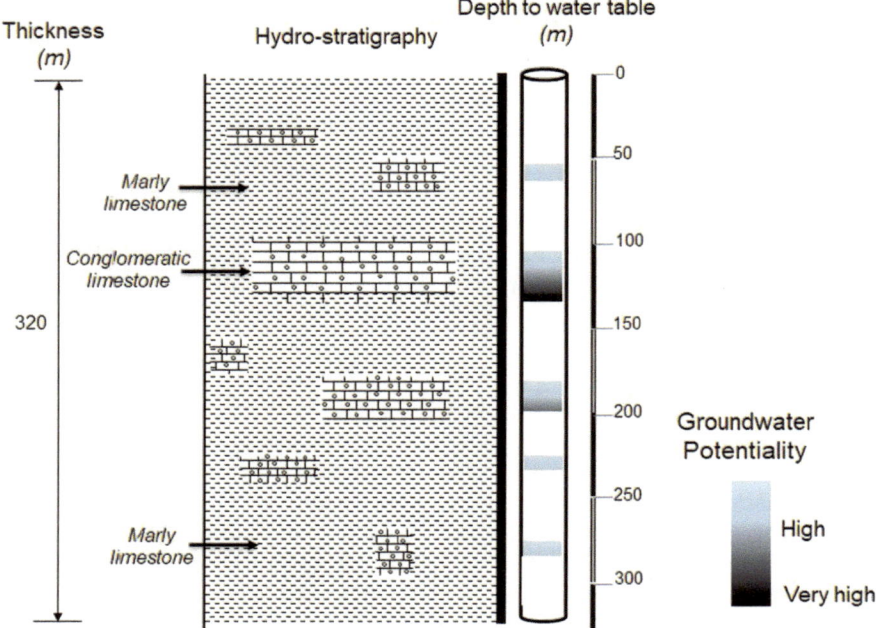

Fig. 8.9 The hydro-stratigraphic sequence of the Neogene Aquifer

in the fractures and karstic conduits at the bottom of the Lutetian rock formation where it is found usually at depths between 500 and 600 m in South Lebanon, while it is found between 250 and 300 m in the Bekaa Plain.

There is obvious facies changes between the Eocene rocks located in South Lebanon and that in the Bekaa Plain (Fig. 8.10). This in turn is reflected on the change in depth and even the discharge from this aquifer.

4. Cenomanian Aquifer:

The Cenomanian Aquifer is the most significant and invested aquifer in Lebanon. This is due to its hydrologic property by including considerable volume of water, in addition to its relatively moderate water table. It also has wide geographic distribution in many Lebanese regions (Chap. 2, Fig. 2.6).

This aquifer, with its large rock succession (~ 700 m), has two potential aquiferous units (i.e. upper and lower) of limestone and dolomitic limestone which are separated by an Aquiclude unit (in the middle) with approximately 60 m of impermeable marl (Fig. 8.11).

Therefore, groundwater is stored in the upper and lower units of the Cenomanian Aquifer, and thus two main groundwater levels are always considered (Fig. 8.11). Upper level (i.e. upper) ranges between 250 and 350 m, while the lower one ranges between 500 and 700 m. Moreover, some perched units may store water at low

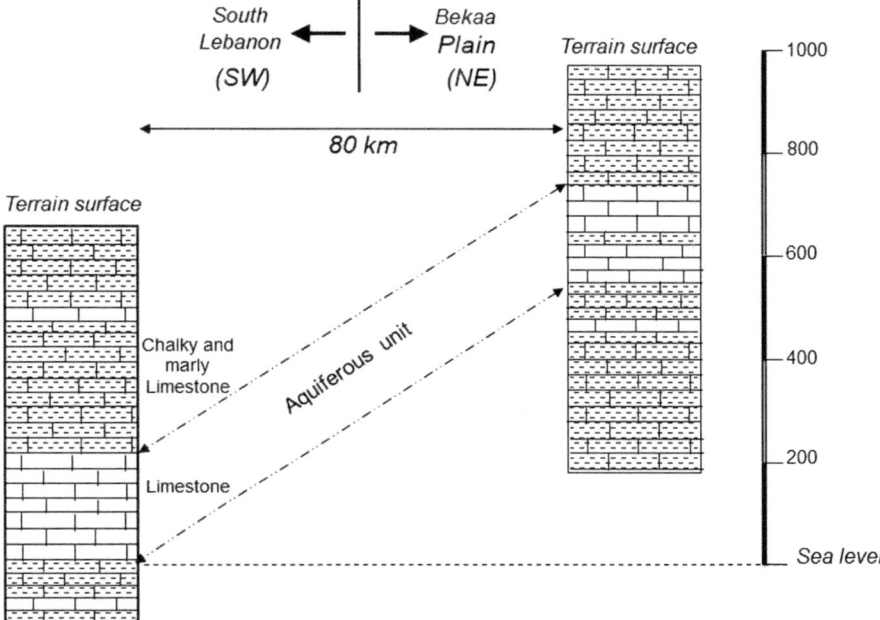

Fig. 8.10 Change facies in the Lutetian rock formation between South Lebanon and the Bekaa Plain

depth (150–200 m), but with little discharge and in many instances it is an intermittent discharge.

5. Upper Aptian Aquifer:

This aquifer is mainly represented by cliff and massive rock exposures. It is largely composed of dolomitic limestone and dolomite where tremendous elongated and large-scale (i.e. few meter long) joints occur from the bottom to the top of this rock unit. Therefore, it is highly porous and permeable and can store water with considerable volume. Hence, the estimated water table varies with wide range between 100 and 400 m.

6. Neocomian-Barremian Aquifer:

This is the unique clastic rock aquifer in the entire Lebanon's rock sequence where sandstone is the main rocks formation with considerable porosity, but the presence of argillaceous and clayey rocks reduces its aquiferous property.

Groundwater in this rock formation is usually with limited volume, and do not have obvious hydro-stratigraphic uniformity. Therefore, the level of groundwater is usually variable and cannot be precisely identified. This makes hydrogeologists consider that the water table of the Neocomian-Barremian Aquifer starts at its top

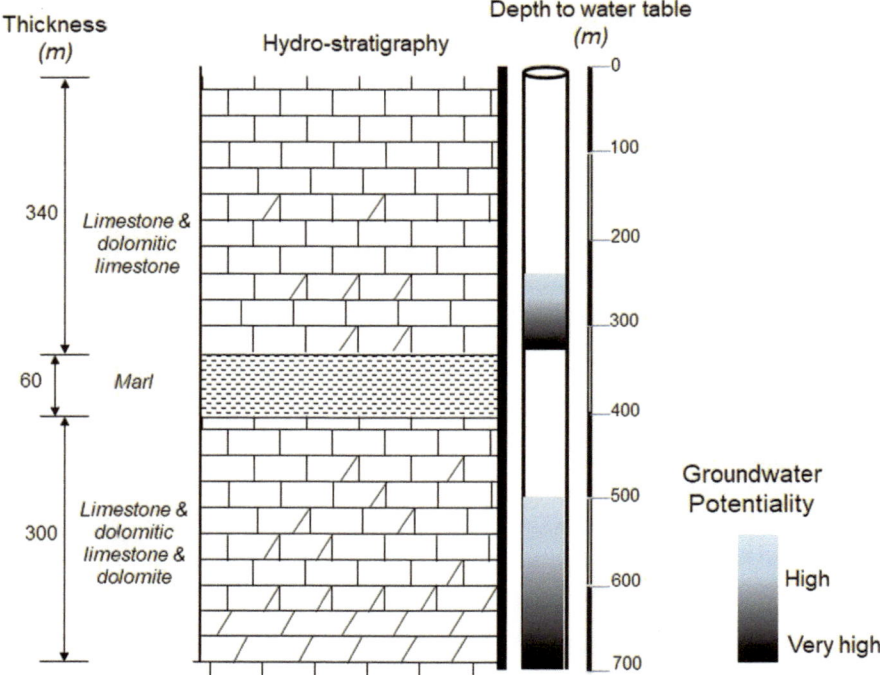

Fig. 8.11 The hydro-stratigraphic sequence of the Cenomanian Aquifer

units moving downward. It, therefore, depends on where the locality of interest is and then by calculating the thickness of the exposed rock that overlying the Neocomian-Barremian rocks.

7. Kimmeridjian Aquifer:

The Kimmeridjian Aquifer is the second significant aquifer in Lebanon after the Cenomanian Aquifer, and it even has better aquiferous properties. However, it has less thickness if compared with the Cenomanian Aquifer and usually it is found at considerable depth which makes it tedious to take a decision of drilling a bore-hole there.

The majority of water storage in the Kimmeridjian Aquifer implies the presence of the secondary porosity among the fractures and karistified carbonate rocks, and in particular within the dolomitic limestone masses. Therefore, these rock masses retain water upon the underlying Aquitard of the Oxfordian rock formation.

The geographic area of this aquifer is mainly in Keserwan Region (Mount-Lebanon) and also along the southern flanks of the Bekaa Plain (Chap. 2, Fig. 2.6) where the latter is located almost in an area with rugged topography and few urbanism. Therefore, the exploitation from this aquifer is limited to Keserwan Region where the depth to groundwater varies between 350 and 600 m.

8. Callovian Aquifer:

As the oldest and lower aquifer in Lebanon, the Callovian Aquifer has a very limited areal extent where it is exposed only in small areas, such as in Naher Al-Kaleb River and some localities near Jezzine. This makes it with little significance as a source of groundwater and it is rarely considered in the hydrogeological studies and in groundwater assessment including water table and discharge.

8.6.2 Groundwater Discharge

The pumping of groundwater from boreholes is always controlled by the amount of stored water, depth of water table and water demand. In this respect, the discharge must be tested primarily when water occurs in the dug borehole, but this is not always the case except for the public wells where pumping test is applied.

The discharge from different aquifer in Lebanon can be summarized as follows:

1. Quaternary Aquifer: The discharge from the Quaternary deposits is varied sharply from one locality to another, but it is in general very low (< 5 l/sec). Therefore, many boreholes which are dug in this aquifer are characterized by intermittent discharge.
2. Neogene aquifer: In the Bekaa Plain, the discharge from the Neogene aquifer is relatively low (10–20 l/sec) if a borehole is dug to about 150 m, but this discharge does not exceed 2 l/sec if the depth of drilling is below 100 m. While, the discharge in Al-Koura Region ranges between 20 and 50 l/sec (Al Ajam 1992).
3. Lutetian Aquifer: The limestone rocks in this rock formation is the groundwater reservoir of this aquifer. Therefore, it differs between diverse regions. This is well reflected on the discharge which is estimated at 20–40 l/sec and 5–20 l/sec in the South and the Bekaa Plain; respectively.
4. Cenomanian Aquifer: The discharge from the Cenomanian Aquifer is high and if a borehole is dug whether in the upper or lower units, the average discharge often ranges between 30 and 50 l/sec.
5. Upper Aptian Aquifer: The exposed cliffs of this rock formation, in many instances, release the contained groundwater. Nevertheless, when this aquifer is totally covered by the overlying rock formations, it then captures groundwater. While the estimated discharge is found between 23 and 60 l/sec according to Al-Ajam (1992).
6. Neocomian-Barremian Aquifer: This aquifer is well characterized by temporary water seeps along the contacts between sandstone and clays. While the most frequent discharge ranges between from 5 to 15 l/sec.
7. Kimmeridjian Aquifer: The discharge from the Kimmeridjian Aquifer is frequently changing even within proximate areas. This is attributed to the development of the sub-surface karstification among this aquifer. Therefore, a discharge between 30 and 40 l/sec is usually found.

8. Callovian Aquifer: The discharge from this aquifer is almost similar to that in the Kimmeridjian Aquifer, but it is often declined very rapidly due to the developed karstification.

8.7   Groundwater Wells

Lebanon can be described as one of the countries with abundant and number of water wells. This is because of the lack to formal controls and the increased demand for water in the absence of public water supply, plus the proximity of dug wells to the localities of water needs (e.g. cultivated lands, etc.). Therefore, all aquiferous rock units in Lebanon are exploited, but private boreholes are almost dug into the shallow aquiferous rock units as this is governed by the relatively cheaper cost of drilling.

8.7.1   Public Wells

Public wells are elaborated by the formal water sector, including mainly MoEW, CDR, and Council for South (CS). While, there are several wells dug by the funding introduced from international entities, such as UNIECF, UNDP, etc. Thus, public wells are usually dug after applying hydrogeological studies and the selection of boreholes is based on scientific approaches. Therefore, well logging and aquifer testing are usually reported.

According to MoEW and UNDP (2014), the data acquired from 841 surveyed wells showed that the total discharge is about 248.7 million m^3/year, which points out that the average discharge from each well is approximately 9.5 l/sec. Some of the public wells are properly equipped and they often under the control of MoEW or by the concerned water establishments. For example, there only 112 out of the 841 wells were equipped with piezometers (MoEW and UNDP 2014). Hence, flow meters and piezometers in newly constructed wells are being recommended, as well as there is also concern to rehabilitate old wells.

Since the public wells are executed by the governmental sector with the purpose of providing water with sufficient amounts to major urban settlements; therefore, these wells are usually dug into the major aquiferous rock units where groundwater is abundant, with a special focus on the Lutetian, Cenomanian and Kimmeridjian aquifers.

8.7.2   Private Wells

Private wells is the most widespread aspect of groundwater sources in lebanon, but many of them do not have define data or information and even most of them are unknown by the formal sector. Hence, most of these wells are dug by individual and rarely a borehole is constructed following a hydrogeological study. This created a chaotic abstraction of groundwater and resulted exhaustion of the aquiferous rock units. Therefore, the MoEW applied new rules for digging boreholes, and therefore licenses become official implement before digging wells.

According to MoEW and UNDP (2014), the official database of private licensed wells was obtained from the MoEW. Thus, 20,537 are officially registered until January 2012. Among the officially registered private wells, only 2888 have exploitation permits and rest number have drilling licenses without exploitation permits. This regulation reduced the number of illegal wells and groundwater abstraction becomes controlled in spite that it did not reach the optimal implementation required.

Yet, the number of illegal boreholes (i.e. the largest number of private wells) has been many times increased in the last few years, even though regulations have been lately adopted by MoEW. For example, the density of boreholes in the suburbs of the capital Beirut has been increased from 500 to 1450 wells/km^2. While, water wells in Hermel Region (a rural area in North Lebanon) has been increased from 8000 to 27,000 wells over the last three decades (i.e. 200 wells/km^2). In this regards, it was estimated that in Lebanon the number of illegal (private) wells is 25 times equals wells dug by the formal water sector.

As the site for drilling private wells is selected chaotically where their locality fits the owner purposes (e.g. close to the farms, water ponds, etc.); therefore, these wells do not follow any hydrogeological understanding on the positioning of the aquiferous rock units, and thus they can be found to be even in Aquitard units.

- Based on the obtained surveys and the available data sets, the distribution of the largest number of private wells with respect to the aquiferous rock formations can be summarized as follows:

- Quaternary = 33%
- Pliocene = 6%
- Vindobanian = 27%
- Lutetian = 11%
- Turonian = 3%
- Cenomanian = 7%
- Lower Aptian = 8%
- Neocomian-Barremian = 5%

8.7.3 Drilling Wells

There are many companies for groundwater drilling in Lebanon. They follow different techniques and tools, and most of them do not apply hydrogeological studies or reports, but in some instances the studies are prepared by hydrogeologists, and then introduced to the drilling companies as guidance while drilling (example Fig. 8.12).

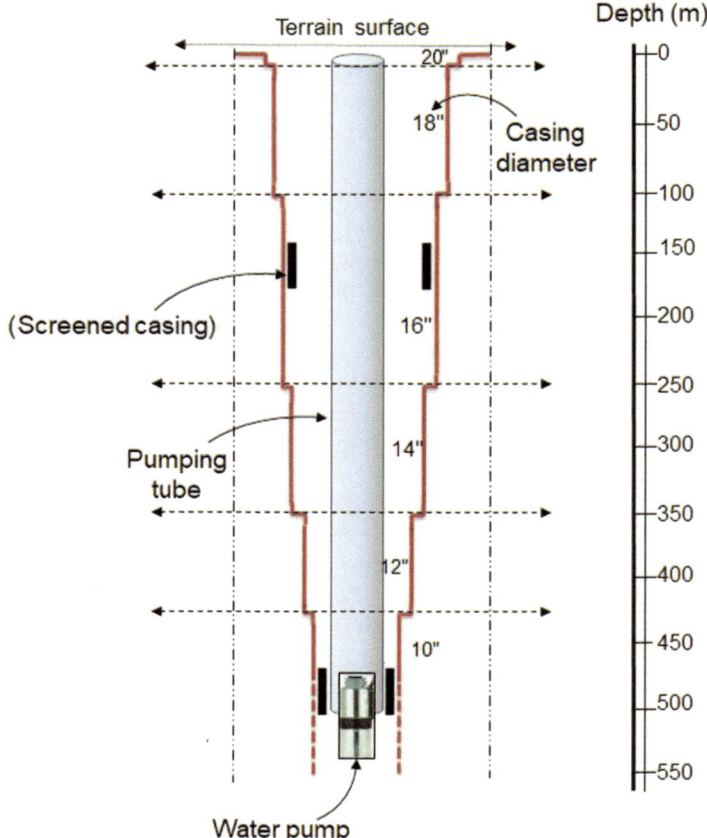

Fig. 8.12 Example of well logging characteristics

Percussion is the most widespread technique for groundwater drilling in Lebanon where a manual approach is followed by using heavy cutting or hammering bit attached to a rope or cable is lowered in the borehole. It is much lea cheaper than the rotary drilling. Therefore, many private wells are dug by this technique, notably when funds are little.

Percussion technique is known in the shallow aquiferous rock units, such as in the Quaternary and Neogene rocks, but it is also found in other rock formations. Thus, it is usually noted that the percussion drilling takes couple of days in the Quaternary Aquifer and this may extend to more than 2 months while drilling in deeper aquifers; especially in hard rock units such as in Lutetian and Cenomanian Aquifers.

Rotary drilling is another technique applied in Lebanon to reach groundwater where a hammer is rotated by the drill stem that provides a passageway through which the drilling fluid is circulated. This technique is more efficient and consumes less time than the percussion one. However, it is much more expensive but lately it become widespread, notably for deep aquifers where groundwater can be reached within a couple of days.

For both techniques of groundwater drilling, there are some constrains usually exist according to the geological characteristics of Lebanon. These constraints imply:

1. Existence of karstic voids (e.g. cavities, galleries, conduits, etc.). Therefore, hammers are usually destroyed and some-time lost into these voids when they are compressed downward with high energy into rocks.
2. Presence of chert pockets and nodules which cause shifting in the drilling hammers due their extreme hardness of these rock bodies.

Chapter Highlights
– There is still an argument about the volume of groundwater in Lebanon. Thus, the estimated values range between 0.5 and 4.84 billion m³/year.
– There are eight aquiferous rock formations in Lebanon where two of them are described excellent aquifers.
– The hydrogeological and hydraulic properties of the aquiferous rock formations have been analyzed as well as numeric estimations were put.
– Only 5.5% of the potential capacity, of the aquiferous rock formations, is occupied by groundwater.
– Based on different thematic influences manipulated in the GIS application, groundwater recharge zones have been identified and it was resulted that 67% of the Lebanese territory is characterized by good recharge property.
– The relationship between groundwater and faults was elaborated. Thus, it was found that the more proximity to faults usually resulted high groundwater potentiality and more productivity in wells.
– The depth and discharge of groundwater were demonstrated according to each aquiferous rock formation.
– Dug wells (private and public) were discussed in this chapter showing the impact of these well on groundwater resources in Lebanon.

References

Abbud M, Aker N (1986) The study of the aquiferous formations of Lebanon through the chemistry of their typical springs. Leb Sci Bull 2(2):5–22
Ahmed F, Andrawis A, Hagaz Y (1984) Landsat model for groundwater exploration in the Nuba Mountans. Sudan Adv Space Res 4(11):123–131
Al Saud M (2010) Mapping potential areas for groundwater storage in Wadi Aurnah Basin, western Arabian Peninsula, using remote sensing and geographic information system techniques. Hydrogeol J 18(6):1481–1495
Al-Ajam N (1992) Aquifers in Lebanon and their characteristics. 1st symposium on the status of water resources in Lebanon (In Arabic), 29–40pp
Bilal A, Ammar O (2002) Rainfall water management using satellite imagery: examples from Syria. Int J Remote Sens 23(2):207–219

Brassington R (1990) Field hydrogeology. Geological Society of London. Professional handbooks series. Open University Press/Milton Keynes/Halsted Press/Wiley, New York, 175p

Canaan G (1992) The hydrogeology of western slopes and coastal plain of Zahrani-Awali region. MSc dissertation, Geology Department, AUB, 85p

Dubertret L (1955) Carte géologique de la Syrie et du Liban au 1/200000me. In: 21 feuilles avec notices explicatrices. Ministère des Travaux Publics. L'imprimerie Catholique, Beyrouth, 74p

Edet AE, Okereke CS, Teme SC, Esu EO (1998) Application of remote sensing data to groundwater exploration: a case study of the Cross River State, southeastern Nigeria. Hydrogeol J Springer 6(3):394–404

FAO (2008) Information system on water and agriculture- aquastat Lebanon. FAO, Rome. Available at: http://www.fao.org/nr/water/aquastat/countries/lebanon/index.stm

Ganapuram S, Kumar G, Krishna I, Kahya E, Demirel M (2008) Mapping of groundwater potential zones in the Musi basin using remote sensing and GIS. Adv Eng Softw 40(7):506518

Gustafsson P (1994) SPOT satellite data for exploration of fractured aquifers in a semi-arid area in Botswana. Hydrogeol J Springer 2(2):9–18

IWMI and USAID (2017) Groundwater governance in Lebanon: the case of Central Beqaa. Groundwater governance in the Arab World. A policy white paper, 36pp

Khadra W (2005) Hydrogeology of the Damour Upper Sannine-Maameltain aquifer. MSc dissertation, Geology Department, AUB, 235p

Khawlie M (1986) Land-use planning for the redevelopment of a disrupted urban center: Beirut, Lebanon. Int J Dev Technol 4:267–281

Khawlie M, Shaban A (1998) Contribution of remote sensing studies to the fractured karstic coastal terrain, Lebanon-Water enhancement or hindrance. Conference on: flow, friction and fracture, AUB Center for Advanced Math. Sciences, Beirut, 1–7 July 1998

Kruseman G, De Ridder N (2000) Analysis and evaluation of pumping test data. Publication 47. International Institute for Land Reclamation and Improvement, Wageningen, 372p

Kumar P, Gopinath G, Seralathan O (2007) Application of remote sensing and GIS for the demarcation of groundwater potential areas of a river basin in Kerala, southwest coast of India. Int J Remote Sens 28(24):5583–5601

MoEW (2010) National water sector strategy. Available at: http://extwprlegs1.fao.org/docs/pdf/leb166572E.pdf

MoEW and UNDP (2014) Assessment of groundwater resources of Lebanon, 88pp

Robinson C, El-Baz F, Singhory V (1999) Subsurface imaging by RADARSAT: comparison with Landsat TM data and implications for groundwater in the Selima area, northwestern Sudan. Remote Sens Abstr 25(3):45–76

Savane I, Goze B, Gwyn H (1996) Etude cartographique et structurale à l'aide des données Landsat de la région d'Odienne [Cartographic and structural study using Landsat in the Ivory Coast]. In: Proceedings of the 26th international symposium on remote sensing of environment, Vancouver, BC, 25–29 March 1996, pp 92–97

Sener E, Davraz A, Ozcelik M (2005) An integration of GIS and remote sensing in groundwater investigations: a case study in Burdur, Turkey. Hydrogeol J 13(5):826–834

Shaban A (2003) Etude de l'hydrogéologie au Liban Occidental: Utilisation de la télédétection. PhD dissertation, Bordeaux 1 Université, 202p

Shaban A (2010) Support of space techniques for groundwater exploration in Lebanon. J Water Resour Prot 5:354–368

Shaban A (2012) A brief review of groundwater resources in coastal Lebanon. In: The INCAM environmental and coastal ecosystem management. CIHEAM, IRD, EU, UK, 213p

Shaban A (2017) Hydrogeological study for Tell Amara, Bekkaa Region. Unpublished technical report (In Arabic), 13p

Shaban A, Dawrich T (2011) The role of sinkholes in groundwater recharge in mountain crests of Lebanon. Environ Hydrol J 19(9):2011

Shaban A, Hamzé M (2017) Shared water resources of Lebanon. Nova, New York, p 150

Shaban A, Khawlie M (2006) Lineament analysis through remote sensing as a contribution to study karstic caves in occidental Lebanon. Revue Photo-interprétation. AGPA Edition. 4:0

Shaban A, Khawlie M, Bou Kheir R, Abdallah C (2001) Assessment of road instability along a typical mountainous road using GIS and aerial photos, Lebanon – eastern Mediterranean. Bull Eng Geol Environ 60:93–101

Shaban A, Khawlie M, Abdallah C (2005) Use of remote sensing and GIS to determine recharge potential zones: the case of occidental Lebanon. Hydrogeol J 14(4):433–443

Shaban A, El-Baz F, Khawlie M (2007) The relation between water-wells productivity and lineaments morphometry: selected zones from Lebanon. Nord Hydrol 38(2):178–201

Tabet C (1978) Geology and hydrogeology of the Baskinta-Sannine area, Central Lebanon. MSc dissertation, Geology Department, AUB, 123p

Teeuw RM (1995) Groundwater exploration using remote sensing and a low-cost geographic information system. Hydrogeol J Springer 3(3):21–30

UNDP (United Nations Development Program) (1970) Liban-Etude des Eaux Souterraines. United Nation, New York

Chapter 9
Challenges on Water Resources

Abstract The concept that Lebanon is characterized by abundant water resources has become an issue of debate. Thus, water supply/demand is always a national problem and water shortage becomes a striking challenge effects several water-related sectors. In Lebanon, the hydrologic regime is not defined yet, while the lack to comprehensive data and records represents one of the principal constraints to identifying the mechanism of the hydrologic system. There are a number of natural and anthropogenic challenges that influence the availability and supply of water resources in Lebanon. From the physical point of view, the climatic variability is the major aspect where the precipitation pattern has been oscillating and it often becomes torrential, while the temperature has been increased at 1.8 °C. The anthropogenic impact has been also exacerbated, and it can be considered with much impact on water resources than the natural challenges. Thus, there is remarkable increase in the population rate estimated at 1.2%; in addition the displaced refugees increased the stress on water supply. Besides, the discharge in rivers and springs has been regressed to more than 50% over the last five decades, and this has been accompanied with a lowering of water table in the major groundwater reservoirs and the pumping rate is also declined to more than 30%. There are many implementations taken to enhance the water sector in Lebanon, but they could not resolve the problem. This chapter will highlight on the existing challenges and how they created stress on surface water and groundwater re-sources in Lebanon.

Keywords Water shortage · Quality deterioration · Increased population · Poor management · Groundwater

9.1 Concepts

There is often debate about the difference between challenges and problems, while it must be clear that challenges are solvable problems, which means there is capability to treat such problems if appropriate implementations are adopted. Also,

A. Shaban, *Water Resources of Lebanon*, World Water Resources 7,
https://doi.org/10.1007/978-3-030-48717-1_9

adaptation can be applied to overcome the impact of challenges. Hence, challenges may occur in ordered phases (e.g. primary, developing, etc.), and this helps making solutions, notably if they are still in the initial phases.

The Global Risk Perception Survey conducted by 900 recognized experts by the World Economic Forum reports that the highest level of societal impact over the next 10 years will be from water crises (WEF 2015).

According to the Environmental Assessment Agency (PBL 2018), water security belongs to three challenges. These are: water scarcity (i.e. water shortages and high demand), water pollution (contaminated water) and water risk (e.g. floods, sea level rise, etc.). These challenges are anticipated to be exacerbated with much impact on human being as a result of increased population, economic development, demand for agricultural and due to climate change.

Understanding of the complexity of water-related challenges and the existence of possible gaps is essential to secure water resources. It would serve as background for the development of sustainable strategies that can adequately reduce risks for the population, economic development, ecosystems, and water associated migration and conflicts (PBL 2018).

Likewise many other vital sectors, water is witnessing challenges in many regions worldwide. However, these challenges are different from one region to another depending on many influencing factors including mainly the physical setting of the region, population growth and even the life style; and the creditability of implementations taken to face them. Therefore, identifying the existed challenges and their elements must be perquisite in order to have an obvious vision on the approaches for reducing and mitigating the impact resulted from challenges and this identification is fundamental element and clue for proposing management solutions which also must be different between regions.

It must be made clear that occurrence of challenges cannot be adopted unless analyses are applied to evidence their occurrence, as well as the degree of impact should be also identified. The analysis of these challenges has a wide variety of approaches and tools depending on their aspect and the region where they exist.

Figure 9.1 shows a framework for the major elements of challenges on water resources. These elements can be summarized as follows:

1. Sensing the problem: It is the initial behavior when people start sensing the unusual behavior in water supply. Thus, water shortage become tangible and obviously occurs. It can be also sensed when poor water quality is noticed, and therefore, the problem of both comes together.
2. Comparison with ordinary: In order to assure the existence of the challenge, people are usually compare the newly existed status with the ordinary (i.e. known status). This can be by comparing water volume, whether the supplied or the stored, with standard volumes, and sometime the supply/demand ratio is accounted.
3. Analysis: This element is often adopted by the scientific boards who apply scientifically-based methods not only to confirm the challenge existence but also to determine its behavior, phase of occurrence and its impact in space and time.

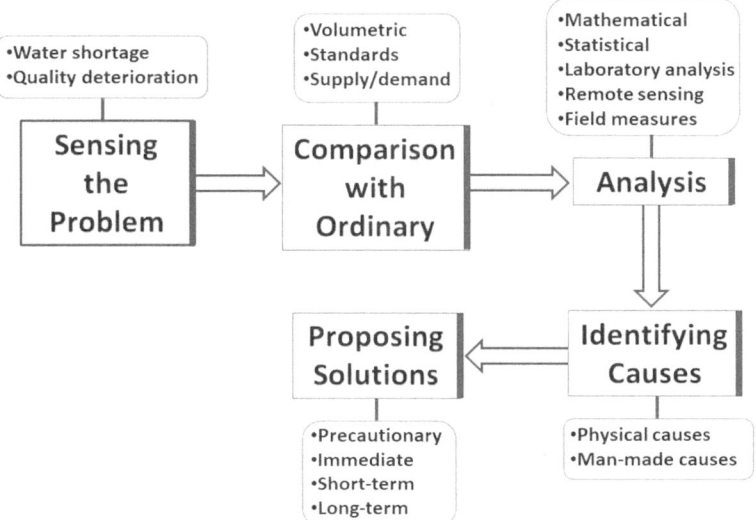

Fig. 9.1 Elements of challenges on water resources

For the creditable analysis, several approaches and tools are considered. This can be by applying mathematical, statistical methods. Also, laboratory analysis is commonly elaborated, notably for water quality analysis. In addition, remotely sensed techniques are utilized to calculate the changing geo-spatial dimensions (e.g. area of water bodies, etc.). Thus, all the above methods are often done along with field surveys.

4. Identifying causes: It is significant to identify the reasons behind the existence of the challenge on water resources. Once the reasons (or factor of influence) are determined, it would be easier to know where and when the treatment must take place. This includes the natural and anthropogenic causes.

5. Proposing solutions: As a result of the analysis and causes identification; however, solutions can be proposed accordingly. This have a wide range of implements which must be taken primarily as precautionary implements. Nevertheless, when the challenge is confirmed, therefore, solutions to be proposed should follow time-frame; in particular, the immediate, short-term and long-term implements.

9.2 Method of Analysis

In order to analyze the geographic and temporal dimensions of challenges on water resources; however, the focus of investigation should be on the water supply/demand balance and also focusing on the main factors impacting water supply where the later should include the physical changing regime of water as well as the interference of human whether by increased water demand or unfavorable behavior with water resources.

According to Shaban (2019a, b), this can be achieved by investigating the time-related origin of challenges which can be classified as:

1. Inherent challenges which are originally occurred since the origin of any region, and then affecting the occurrence and the hydrologic regime of water resources, such as the altitude, topography, geologic setting, etc.
2. Intervened challenges which have been occurred lately and influenced water volume and quality, such as the increased population and increased agricultural land with more need for water, etc.

Therefore, empirical assessment and optimal analysis, for water resources under stress, can be applied for both time-related origin challenges, and this helps identifying actions needed to reduce the impact of these challenges. Hence, the phases for challenges analysis are adapted from Shaban (2014) as illustrated in Table 9.1.

These phases imply mainly the preparation of data and information required; hence, long-term and comprehensive datasets can be utilized for further analysis including mapping and establishment of trends and projections by using advanced statistical methods. This can be supported by using *in-situ* measures and other measures extracted from remote sensing and obtained in the laboratory analysis. Therefore, field surveys can be applied at different stages of work. In addition, there must be benefit from the successful stories and studies applied at different scales whether on the national or international levels.

9.2.1 Data Preparation

As a required pre-processing step, data preparation is the initial phase in which data and information are gathered from different sources for further data clearance, transformation and analysis. In this respect, the author depended on his research studies done on water in Lebanon where a number of them focused on the existing challenges (Shaban 2014, 2018, 2019a, b). This was not limited to the topic challenges analysis but also for general assessment and analysis of water resources and different hydrologic regimes.

The collection of data and records is always faced by many constraints. In particular, the climatic and hydrologic data records are often unavailable like the case in Lebanon. While these data and records (if exist) are found with limited time period or just for small areas. In addition, most of the data collected were different when they were collected from different sources; therefore, contradictory outcomes are also resulted. Therefore, the unavailability of data and the non-uniform distribution of measuring stations is always a problem.

The author, in previous studies, has collected comprehensive data from several sources (as they were almost mentioned in Chap. 2, Sect. 2.1). However, the

Table 9.1 Phases for analyzing challenges on water resources

Phase	Description	Expected outcomes
Inventory on data and information	Previous studies, researches, etc. Implemented plans and projects Existing strategies and policies Long-term climatic and hydrologic records	Preparing inventory on water resources in the country
Data analysis and interpretation	Maps analysis (including illustrators) Obtained case studies on water resources Statistical analysis of measures (e.g. water consumption population size, etc.)	Identifying the current measures and their trends
Use of advanced tools	In-situ measurements Advanced laboratory techniques Remote sensing and geo-spatial products	New tools and techniques will give accurate results
Field verification	Field investigations (mainly those obtained by the author) Results data from previous field observations and investigations	Verifying obtained resulted and adding filed measures
Utilizing from lessons learned	Benefit from successful studies and projects obtained regionally and globally Identifying useful tools used in water resources assessment Inducing reasons of failure in water sectors	Identifying elements of success or failure and benefit from these element for better management approaches

Adapted from Shaban (2014)

collected data were organized, sorted and the gaps were identified and then filled either by the statistical interpolation or from the remotely sensed products. Therefore, the sources of collected and prepared data on climate and hydrology, which are necessary for water resources assessment, and for the related analysis of the existing challenges, can be summarized as follows:

1. Local hydro-climatic data:

   - CNRS-L/CESBIO: These belong to three meteorological stations fixed by CNRS-L and Centre deludes Spatiales de la Biosphère (CNRS-L/CESBIO 2019).
   - Litani River Authority: This includes hydrologic measures, notably for rivers and springs discharge (LRA 2017).
   - Lebanese Agricultural Research Institute (LARI). Climatic Data. Monthly Bulletin. Department of Irrigation and Agro-meteorology (LARI 2017).
   - General Directorate of Civil Aviation - Direction Générale de l'Aviation Civile (GDAC 1999).

- Climatic Atlas of Lebanon: Atlas Climatique du Liban (CAL 1973, 1982).
- Climatic data from different local sources: This includes studies, theses, technical reports, etc. (example: Ghaddar 2003). In addition, climatic measurements provided by Lebanese institutes, such as the American University of Beirut (AUB).

2. Global climatic data:

   - TRMM: Tropical Rainfall Mapping Mission: This is a remotely sensed system based on the radar data. It retrieves rainfall datasets on daily basis for different regions worldwide (TRMM 2014).
   - CHIRPS: Climate Hazards group Infrared Precipitation with Stations. Its algorithm built on ground-climatic measures incorporates with satellite information. It provides daily and monthly data (Funk et al. 2015).
   - NOAA: climatic data system - National Oceanographic Data Center (NOAA 2013).
   - Miscellaneous data sources: Meteorological datasets retrieved from obtained by projections and scenarios elaborated for global hydro-climate analysis. A typical example is the Intergovernmental Panel for Climate Change, Working Group I (IPCC 2010).

3. Thematic data:

   - In addition to the climatic and hydrologic measures, there are also thematic illustrations, maps, geo-spatial information and studies on water resources were prepared from previous studies obtained by the author (mentioned in the reference list of the book chapter) and from another sources. This include mainly maps on: topography, geology, drainage systems, hydrogeology and DEMs. In addition to a number of satellite images with different spatial and temporal resolution.

9.2.2 Data Analysis

As a first step for data analysis, the collected data will be reviewed and investigated in the view of identifying the factors resulting in challenges on water resources. Consequently, comparison and change identification will be the second step before reaching up with conclusions. Figure 9.2 summarizes these steps and their influencing items.

1. Review and data investigation: This includes calculating water entering the hydrologic system and loses (& consumption), and then water allocation by sector and the applied management approaches plus the strategies and policies adopted.
2. Comparative analysis: It represents the majority of challenges analysis since it implies comparing volumetric measures for surface and groundwater resources

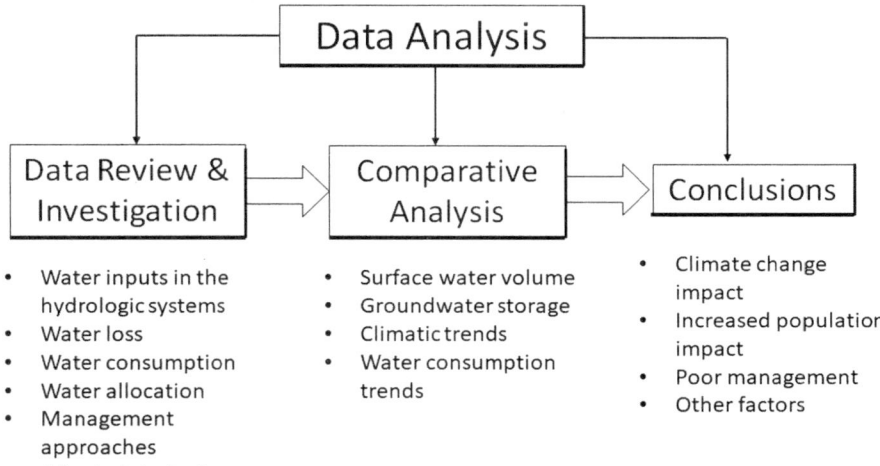

Fig. 9.2 Data analysis for the challenges on water resources

based on statistical analysis for data with long time series. This also involves the analysis and comparison of climatic elements with a special emphasis on precipitation and temperature. The comparison should be applied also to water consumption and the changing rate for the per capita must be determined.

3. Conclusion: This is the resultant of data investigation and the applied comparison where it specifies with numeric illustrations the impact of climate change on different sector and certainly on water sector, increased population and their newly requirements of water as well as the impact of the applied management approaches and their success or failure as well as any other factors that might be resulted.

9.3   Natural Challenges

Generally, there is a contradictory in defining the aspects of challenges on water resources in that whether they belong to natural or man-made factors or the combination of them. In this view, both aspects of challenges occur in Lebanon and none could identify which is the most effective and even the degree of impact by each of them is still undetermined (Shaban 2019a, b).

It can be said that the natural (physical) challenges often occur, and they may act as creeping/or immediate impact on water resources and the related disciplines. They are usually attributed to climatic conditions and the other physical factor are often ignored. Physical challenges cover wide geographic areas and controlled by natural factors existed there. Also, dealing (i.e. treating, solving, etc.) with physical challenges is always inaccessible or sometimes limited, but there can be adaptation measures applied to face these challenges.

9.3.1 Topography and Geology

The combination of topography and geology of Lebanon created natural conditions that act significantly on the regime of water resources. Even though, the topography of Lebanon, with its mountainous nature, plays a role in constructing a meteorological barrier that resulting in considerable precipitation rate, yet this topography is also negatively controlling surface water regime. This is also the case for the structural geology which is sometimes play a positive influencer, but it also hinders the regularity of groundwater flow and storage. Therefore, the negative influencers on water resources in Lebanon can be concluded as follows:

1. Terrain slope: The steepness of the terrain surfaces in Lebanon is an acting topographic factors on surface water flow regime. Thus, the majority of slope gradient is almost high and the average slope gradient exceeding 25 m/km over most of the Lebanese mountain chains (Shaban 2019a, b). It is calculated as:

   – Mount-Lebanon: 120 and 160 m/km towards sea side and the Bekaa Plain; respectively.
   – Anti-Lebanon: 60 and 80 m/km towards the Bekaa Plain and Syria; respectively.
   – Selected elevated areas: (132 m/km in Jabal Sannine, 171 m/km in Jabal Hermoun; 86 m/km in Jabal Akroum).

   Generally the steep sloping terrain makes surface water flow energy higher whether along the valleys or even on irregular surfaces. This is well pronounced in Lebanon where the rapidity of surface water flow loses a large part of this water into the sea or to the neighboring countries before any proper exploitation. In addition, the rapid water flow will act in reducing the percolation rate of surface water into the underlying rocks.

2. Rock deformations: The structural geology play a binary process in controlling the groundwater regime. As it was previously mentioned (In Chap. 8) that rock deformations, and notably the fracture systems, serve in increasing the recharge rate and in capturing groundwater along faults; nevertheless, these deformations can be also natural challenges. This is the case in Lebanon where rock deformations, and specifically faults which span from the mountainous regions to the sea, represent fragile zones, where they transport large amounts of groundwater into the sea as a huge water loss (Khawlie and Shaban 1998; Shaban et al. 2005).

3. Sub-surface karst: Similarly to the case of faults, the sub-surface karst are sometimes penetrating into very deep and unreachable aquifers like the case of Lebanon (Khawlie and Shaban 1998; Shaban 2003, 2010).

Moreover, sub-surface karst (likewise faults) also contributes in discharging groundwater into the sea where sub-marine springs occur and they are still considered as water loss without any benefit (Shaban et al. 2005, 2017).

9.3.2   *Meteorological Conditions*

It is well known, as mentioned previously in Chap. 2, that the climate of Lebanon is a favorable whether as it is a mild climate or in providing the country with plenty of water from precipitation. Recently, the changing climatic conditions have been occurred and became challenging, notably by influencing water resources. It is a global challenge that strikes many region of the World.

However, Lebanon is the 78th least vulnerable country and the 59th least ready country for climate change impact (MoFA 2018). Whereas ND-Gain Index (GAIN 2017), reported that the climate vulnerability of Lebanon has a ranking of 106 out of 181 countries in the (where ranking 1 is the least vulnerable).

In particular, the changing precipitation and temperature are the most significant factors, and they can be well analyzed if long-term datasets are prepared accordingly.

1. Precipitation: This includes liquid and solid precipitation (snow) which are utmost significant water sources in Lebanon and they can be considered as plenty. However, both have been subjected to changes over time and this must be normal if the weather periodicities (cycles) are accounted.

   – Rainfall: Based on the analysis of long-term series of rainfall records which were collected from several sources as mentioned in Chap. 2 and in Sect. 9.2.1 in this chapter; therefore, simplified illustrations have been applied for 37 stations in Lebanon separately, and then the average rainfall rate for the entire Lebanon was also elaborated (Fig. 9.3).

An ordinary oscillation in the rainfall rate has been clearly noticed in Fig. 9.3 where a decrease of approximately 12–15 mm has been noticed in the trend of the rainfall rate over the (about) last seven decades (1950–2018). This change cannot be considered as a very remarkable, but relatively moderate. However, it conflicts with beliefs and estimations that rainfall witnessed sharp changes in Lebanon towards a decrease in the volume of water.

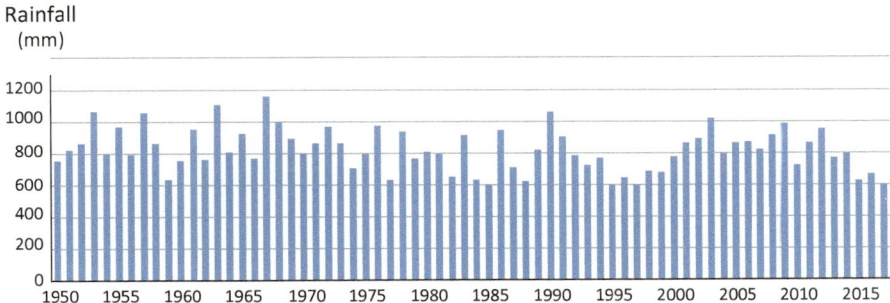

Fig. 9.3 Average annual rainfall for Lebanon (1950–2018)

These findings accorded with the analysis obtained by CNRS-L (2015) where a decrease in rainfall rate have been reported over 52% of Lebanon's territory beside 48% increase. In other words, the analysis obtained CNRS-L point out to noticeable and moderate change in the amount of water from rainfall, but not very much.

– Snow: A detailed discussion in Chap. 5 was on snowpack, its areal extent and melting/accumulation regime and periodicity. However, the changing elements in snowpack were not diagnosed, except the mean monthly snow cover areas was plotted for several years (2000–2018) as shown in Fig. 5.3. Therefore, no obvious changes have been noticed in the snow cover area over the investigated time period.

– Nevertheless, it is observable that the duration of snowpack melting has become more rapid. In other words, the accumulated snowpack is being disappeared within short duration than before. For this reason, the duration of snowpack melt was illustrated in Fig. 9.4. It represents the time period between the first snowfall and till snow has been totally melted (except couple of square kilometers for the entire Lebanon).

Results show that the period for snowpack melting was approximately 3 months between 1989 and 2000, while this period became 2 months after 2010.

2. Temperature:

The analyzed time series for temperature shows a clear changing in the registered maximum and minimum temperatures, where the maximum is increasing toward higher values and the minimum is increasing towards lower values. This in turn created widening in the difference between both values, and this widening influenced many sectors including mainly water and agriculture (Darwich et al. 2014).

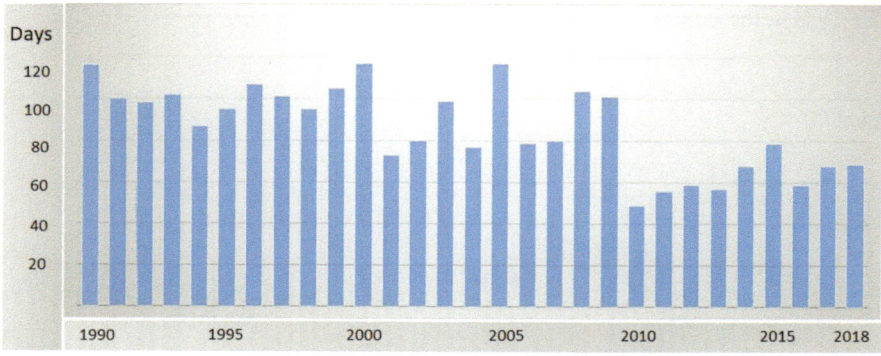

Fig. 9.4 Melting duration of snowpack in Lebanon

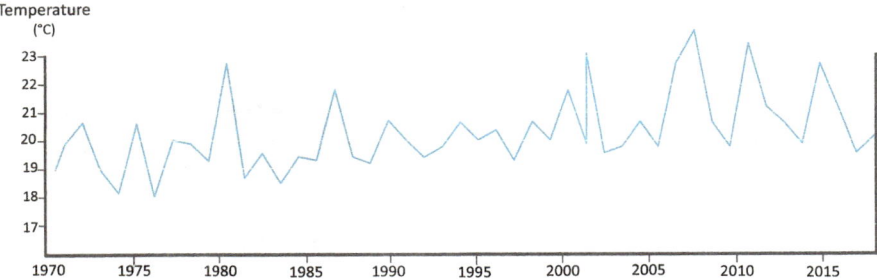

Fig. 9.5 Mean annual temperature of Lebanon (1971–2018)

The increasing trend in the average annual temperature started since 1970s (Fig. 9.5). In addition, sharp oscillating temperature becomes dominant atmospheric phenomenon. Therefore, remarkable temperature increase was estimated at 1.8 °C over about five decades. In this respect, there are many estimates were put, such as the increased temperature of 0.6–2.1 has been reported by Bou Zeid and El-Fadel (2002) and 0.6–2.1, while an increase in annual mean temperature of 0.11 °C per decade was mentioned by USAID (2016).

Even thought, there is no tangible change in water volume entering the hydrologic system of Lebanon's territory, but there are many aspects of climate change and variability influencing this system. These aspects were mentioned in several studies done by the author (Shaban 2011, 2014, 2018, 2019a, b), and they can be summarized in Table 9.2.

9.3.3   Hydrological Elements

Perhaps the oscillating climatic conditions in Lebanon is a major factor influencing the volume and regime of water resources, but this fact remains without any evidence, notably that the increased demand for water is directly governed by the population size and even the ethical behavior of consumers. This makes it necessary to analysis the hydrologic elements for surface water and groundwater resources which are obviously declined.

1. Discharge in rivers and springs: It is well known that all rivers and springs in Lebanon are under decline in the water discharge. This has been well noticed since three decades where most of the Lebanese rivers became totally dry in summer and fall seasons, and some other rivers remain without water for more than 9 months a year. This also created contradictory in defining the exact number of Lebanese rivers. In addition, there are more than 60% of the springs in Lebanon have been disappeared (Shaban 2003, 2011).

Table 9.2 Major aspects of changes and influencers of climate of Lebanon

Element of climate change	Description	Hydrologic impact
Torrential rainfall	Rainfall patterns became torrential and thus wet storm strike Lebanon several times every year	Increased the number and the spatial distribution of floods Reducing recharge rate Increasing water loss to the sea
Increased temperature	According to the elaborated statistical analysis for the mean annual temperature, there is about 1.8 °C increase over 5 decade for the entire Lebanon	Increased evapotranspiration rate Accelerating the melting of snowpack due to increased sublimation Water will face a reduction of 6–8% of the total volume of water resources with the increase in 1 °C (SNC 2011)
Higher snow melt rate	Before 2000: the rate of snowpack melt averages over 97 days. It has been changed to 86 days between 2000–2010 and then 64 days between 2010–2018	Less infiltration rate and fewer groundwater recharge Increased run-off and sometimes water in rivers overflows
Seasons shifting	There is obvious shifting (i.e. time displacement) in weather seasons. It was estimated at 20–25 days.	This indirectly reflected on water resources and the irrigation scheduling processes and irregular water consumption
Abrupt temperate change	Tangible temperature change (>10 °C & e.g. one-day extreme rainfall events) in very short time (e.g. even in the same day)	Turbulence in the hydrologic regime that influences agriculture and then water consumption
Increased dust waves	Increase in frequency of dust waves which derived from the deserts of the Arabian Peninsula and the Western desert	Rising the degree of dryness and then increased the rate of water consumption
Periodicity	The calculated Emberger aridity index showed diverse periodicity that ranges between 2–21 years for the entire Lebanon	This reflects the micro-climate impact and then the imbalanced water distribution

There are several unfavorable works observed on rivers and springs which can be, in addition to the climatic oscillations, the reason behind the decreased discharge in these water resources. These implements can be summarized as follows:

– Direct water pumping from rivers and springs (example Fig. 9.6) without any control.
– Conveying rivers' and springs'water into canals for irrigation purposes.
– Water abstraction whether from the recharge zones, which feed directly rivers and springs, or from recharge zones for groundwater.

The availability of datasets for the discharge from river and springs enabled applying a detailed statistical analysis for a number of these resources on monthly basis (example: Telesca et al. 2013). These datasets, which were provided by the

Fig. 9.6 Direct water pumping from Litani River, Bekaa Plain

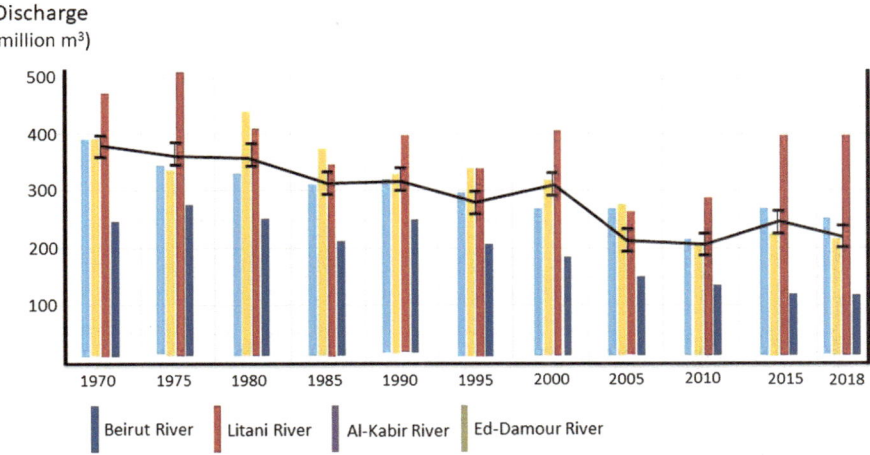

Fig. 9.7 Discharge of selective rivers in Lebanon (1970–2018)

LRA, were also analyzed on the mean annual basis for time series. Figure 9.7 shows the discharge trending in selective rivers and springs for the time period between 1970 and 2018. The illustrated trends show obvious regression in the discharge whether for rivers or springs.

Discharge
(million m³)

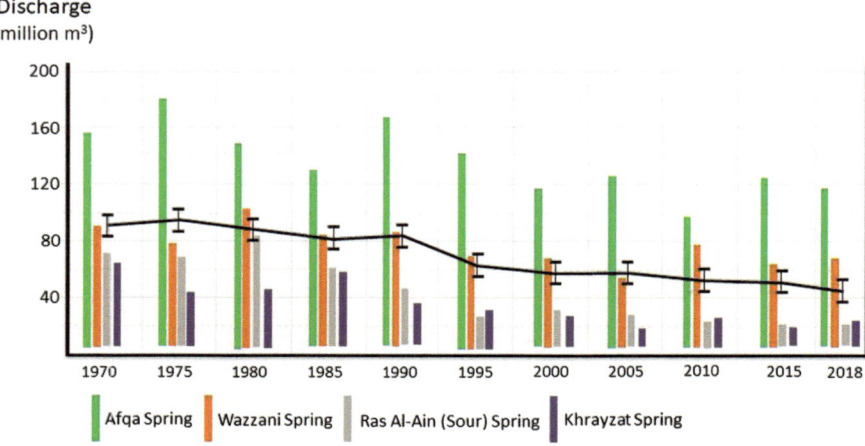

Fig. 9.8 Discharge of selective springs in Lebanon (1970–2018)

The estimated decrease in the discharge of the selected rivers is about 38–45% over (approximately) the last five decades. For some rivers, the decrease exceeded 55% like the case of Beirut River. The decrease in rivers' discharge is a challenge, and it is exacerbated by the intermittency of the discharge over at least 6 months a year.

For springs, it is deduced that the decrease in their discharge is about 38–40% (Fig. 9.8). It is also noticed that some springs have acute discharge decrease (e.g. Ras Al-Ain, Sour and Khrayzat springs) while others show little decrease (e.g. Afqa and Wazzani springs). The reason behind this contradictory can be attributed to the fact that urban activities are well pronounced in the catchments or feeding zones for springs with higher discharge decrease, while little decrease in this this charge is found in springs with minimal urban activities (Fig. 9.8).

2. Groundwater level and discharge:

Groundwater has almost similar bad status as much as that of surface water where it has been influenced by changing natural conditions and more specifically the oscillating climate. Thus, changes groundwater includes the discharge and water table level. All investigated or surveyed wells show regression in the pumping rates where some of them become totally dry. While others just revealed lowering in the water table.

Yet, the exact factor influencing this regression in groundwater is still undetermined and it is always attributed to climate change with less focus on human turn. No doubt that climate change has a role, in this respect and specifically from the existed torrential rainfall pattern which decelerated the recharge process of groundwater.

To verify groundwater behavior, field investigation on drilled water wells has been carried out by the author (Shaban 2012). This has been interrelated with recently applied investigation in order to reach most precise results.

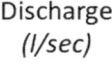

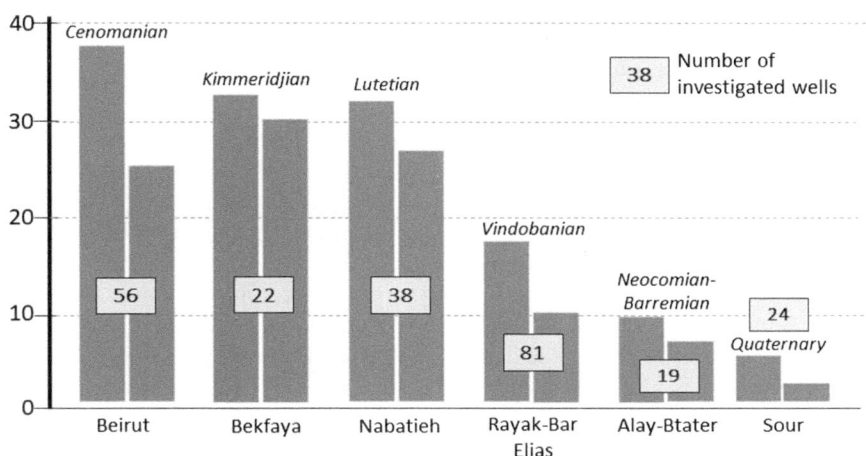

Fig. 9.9 Wells'discharge from selective regions and aquiferous rock formation (between 1984 and 2017 including several time gaps)

The discharge from 240 water wells, located in different Lebanese regions, were investigated where they belong diverse types of aquiferous rock formations including the alluvial deposits, carbonate and clastic rock masses (Fig. 9.9). The discharge of these wells were compared over long time period starting since 1984 until 2017 (Fig. 9.9). However, the measurements over this time interval was for similar months for all wells, but these measurements were not continued and obtained for different time period depending on the availability of collecting data from these wells (e.g. between June and July of the years 1985, 1988, 1994, 1998, 2001, 2006, 2017).

The obtained results show decrease in the amount of water pumped in these wells (discharge). Thus, the decrease was: 33%, 13%, 13%, 44%, 19% and 60% for the Cenomanian, Kimmeridjian, Lutetian, Vindobanian, Neocomian-Barremian and Quaternary aquifers; respectively. This means that the average decrease rate in the discharge from water wells exceeding 30% over about three decades.

This decline in the discharge in wells is usually accompanied with lowering in the water table. However, no precise measuring has been put for the value of drawdown in water levels of wells, but it has become a wide knowledge that the water table gets deeper. In this respect, Shaban (2019a, b), estimated that the drawdown in the major two aquifers of Lebanon as 20–25 m in the Cenomanian aquifer and 8–12 m in the Kimmeridjian aquifer.

3. Saltwater intrusion:

The geology of Lebanon as it is characterized by the existence of fractured and karistified carbonate rocks along the coastal stretch, as well as the uncontrolled and excessive pumping of groundwater, lead to saltwater seepages from the sea into the coastal aquifers. This became widespread hydrologic phenomenon since early 1990s.

Added to the expected sea level rise in Lebanon, which is expected between 30–60 cm by the next 30 years (MoFA 2018), the majority of saltwater seeps into land aquifers includes mainly the flow along hydrologic routes (i.e. faults and karstic conduits), as well as seeps through fissures when the latter extend between land and sea. Therefore, faults and karstic conduits transport salt-water for relatively long distances (5–10 km) on-land. In this respect, several faults and karstic conduits were routes that transport groundwater into the sea as sub-marine springs, but when water table drawdown due to over pumping, these hydrologic routes began working inversely and transporting saltwater in coastal aquifers (Shaban et al. 2005).

Then, it was reported that salinity levels in the coastal groundwater of Lebanon due to intrusion of saltwater has increased 200 fold since 1960s where it has been increased from 100 mg/l to 20,000 mg/l (Lababidi et al. 1987; Khair et al. 1994). Also, it has been indicated that chloride concentrations from 125 randomly sampled wells had risen from 340 mg/l in 1972 to over 4200 mg/l in 1985 (Khair 1992).

A study was applied (1999–2002) for a typical coastal region in Lebanon which is located between Choueifat, Jiye and Rmaiyleh (24 km). Then, it was concluded that this region is subjected to saltwater intrusion, and the salinity rates oscillated between 0.7dS/m and 5.5dS/m (Bakalowicz 2009).

The salinity rate is much more excessive adjacent to the urbanized coastal cities, notably the capital Beirut where salinity levels were reported to be over 5000 mg/l in some of the surveyed public and private wells (Saadeh 2008). While, recently sampled groundwater from boreholes in Beirut showed that the Total Dissolved Solids (TDS) with thousands of milligrams per liter; and sometimes it is equivalent to that of seawater, i.e. about 37,500 mg/l (Saadeh and Wakim 2017).

4. Other resources:

Other than the mentioned hydrologic elements, there are many other hydrologic elements which have been lately changed and became challenging elements influencing water resources. The most important ones can be summarized as follows:

– Many wetlands in Lebanon are losing their characteristics as a water body (mentioned in Chap. 7, Sect. 7.2.2). A good example is the RAMASAR wetland of Cliffs of Ras Ech-Chekkaa, which has been totally dried (Shaban et al. 2016).

- Many lakes and reservoirs are subjected to decrease in their dimensions and then reduction in water capacity. Thus, the Qaraaoun Reservoir, the largest of its type in Lebanon, is an example. Therefore, the processed Aster and Landsat 7 satellite images showed that the Reservoir reveals several dimension where the average areal extent has been changed from 11.63 km² in 1999 to 8.56 km² in 2018.
- Water loss in Lebanon is dominant and it represents a major physical challenges on water resources. This includes two major aspects of water loss: (a) water loss into the sea whether as surface run-off or as groundwater seeps into sea water (i.e. sub-marine springs), (b) water loss into deep karstic aquifer where groundwater moves into undefined conduits and shafts.

9.3.4 Trans-Boundary Water

Other than the coastal stretch along the Mediterranean side, Lebanon has approximately 559 km border; therefore, a large volume of surface and groundwater are shared with neighboring countries without any tangible benefit from this water. In this respect, Shaban and Hamzé (2017) stated that if Lebanon properly utilized about 50% of its shared water resources; therefore, 616 million m³ will be added to its water budget which is equivalent to about 154 m³ capita per year.

About 27.5% of Lebanon's area constitutes basins for Transboundary Rivers with the riparian regions. There are two international rivers (i.e. Orontes River with Syria and Turkey and Jordan River with Syria, Jordan and the Palestinian Territories) that have major watercourses span from Lebanon, while the third river (El-Kabir River) represents the national border between Lebanon and Syria. Hence, the average annual discharge from these rivers (from the Lebanese side) is approximately 867 million m³per year (Shaban and Hamzé 2017).

Also, there are about 2631 km² of Lebanon' area forming shared groundwater basins with the neighboring regions. These sub-surface basins represent potential rock formations for groundwater storage which was estimated, by Shaban and Hamzé (2017), at approximately 365 million m³ of water.

9.4 Man-Mad Challenges

It is not exaggeration to say that the impact of human interference on water resources in Lebanon is much more harmful than the natural influencers. This concept can be supported when the natural challenges are analyzed and compared with the degradation level in water sector in Lebanon.

Yet, the majority of man-made challenges on water resources in Lebanon is directly attributed to the increased population size and the related increased water demands, whereas this is not very accurate if the behavior of water consumers is neglected. Therefore, the degradation in water sector in Lebanon has two major pillars including the quantity and quality.

Unlike the natural challenges, man-made challenges on water resources can be treated, reduced or even stopped if wise management approaches and appropriate implementations are taken by consumers and this needs ethical and economic controls on water use.

9.4.1 Population Growth

Generally, the increased population is directly reflected on water demand, and thus water supply must be regulated accordingly. Nevertheless, if water supply remains with no adaptation measures while the population in increasing; therefore, water deficit inevitably occurs, such as the case of Lebanon.

The population size in Lebanon increased by about 1.1% and 1.3% as put by Index-Mundi (2017) and WB (2017); respectively. Therefore, the average (1.2%) and the population of Lebanon (~4.3 million people), hence, about 50.000 people will be added to the total population.

Based on a socioeconomic survey that has been carried out by the author to identify the major influencing elements water resources, it was resulted that the average demand per capita is approximately 220 m^3/capita/year (Shaban 2011), this mean there will be 9.5 million cubic meter of water is additionally required.

The above population size has been reported until 2013 (WB 2017) after which a large number of refugees has been displace to Lebanon due to the political conflicts. This number is often under debate and it ranges between 1.5 and 2.5 million people. Therefore, about two million people supposed to be added to the Lebanese population and the total number will approximately 6.3 million people.

Figure 9.10 shows that the population of Lebanon was about 3.7 million people in 2003, and then the total water needs was about 829 million m^3. This was slightly increased till 2013 when the population became 4.3 million and total water demand reached about 930 million m^3 (Fig. 9.10). Thereby, the increased population (about two million refugees) between 2013 and 2017 has raised the total water demand up to about 1390 million m^3. This means that about 460 million m^3 has been added to water demand.

There must be made clear that the increase in water demand is not only attributed to the population increase but also on the newly and civilized requirements for water as well as the unaccounted for water which have been exacerbated lately. This includes a variety of consumption aspects which were not consumed before, such as (but not limited): car washing, touristic resorts, swimming pools, water for construction, firefighting and leakage due to old infrastructure.

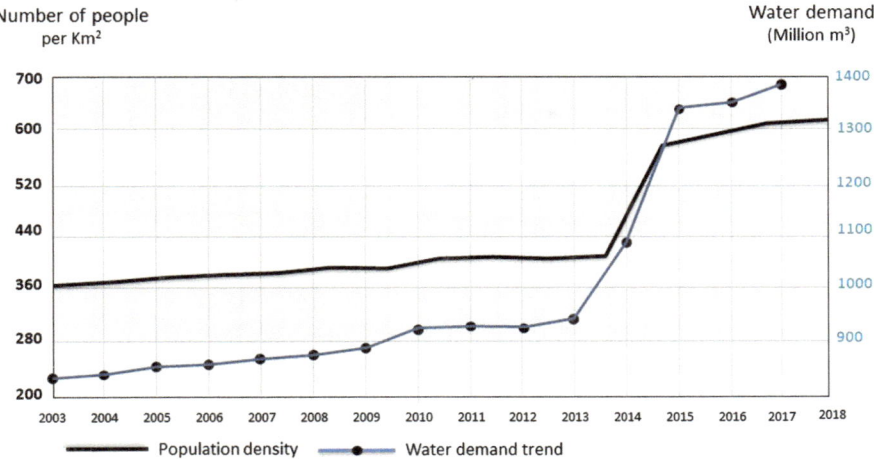

Fig. 9.10 Relationship between population growth and water demand in Lebanon

These recent water requirements occupy integral part of water consumption in Lebanon, notably in the lacking of legal controls and water metering implements. Therefore, new water requirement can be estimated at about 30% of the normal water demand which exceeds 65 m^3/capita/year.

9.4.2   Quality Deterioration

More than 50% of the Lebanese water resources is contaminated (Shaban 2014). Hence, water quality deterioration is a major striking challenge directly touching human life. Hence, sources of pollution are widespread in Lebanon, and they are mainly due to the uncontrolled disposal of liquid and solid wastes including industrial, municipal and even agronomical wastes. This environmental situation put an increasing pressure on the physiochemical and biological contamination of surface and groundwater resources, and thus the quality of water became severe and exceeded several times the international norms.

Implementations for waste disposal management have been undertaken and several national and international projects were applied. In addition, filed campaigns, capacity building, inter-ministerial committees and business plans were established to identify the required measures and secure water quality and the existing ecosystems. However, no improvement has been reached yet to alleviate the pollution, notably on water resources.

There are several studies obtained to assess water quality and to identify sources of pollution in Lebanon, but these studies were either applied to pilot areas or for specific water system (e.g. river, spring, aquifer, etc.), or even for one type of water contamination (e.g. biological).

Different examples can be introduced to provide a comprehensive figure on water quality deterioration in Lebanon as follows:

1. Analyzed water samples from the Litani River obtained by Nemeh and Haidar (2018) were compared with WHO (2006) show the following selective results:

   – Nitrite (NO_2) 19 ppm (max. 0.1 ppm)
   – Chromium (Cr^{3+}) 0.27 (max. 0.05 ppm)
   – Staphylococcus 8750 (0 in 100 ml)
   – Total coliform 183,000 (0 in 100 ml)
   – Fecal coliform 180,000 (0 in 250 ml).

2. The chemical and bacteriological analysis of water in the Qaraaoun Reservoir, as obtained by Fadel et al. 2014, 2015, 2016 shows the following selective results:

   – Chromium (Cr) 0.02 mg/l (WHO max. 0.05 mg/l)
   – Zinc (Zn) 0.09 mg/l (WHO max. 3 mg/l)
   – Copper (Cu) 0.019 mg/l (WHO max. 2 mg/l)
   – Cylindrospermopsin toxin 1.7 µg/L (WHO guidelines 0.7 µg/l)
   – Cyanobacteria up to 200 µg/l (WHO limits 10 µg/l)
   – Carlson trophic state index 66 to 84 (CTSI max. 40).

3. Groundwater analysis in several water wells located in the Bekaa Plain showed that Nitrate (NO_3) concentration exceeded 300 mg L-1 (Darwich et al. 2008).

4. The analysis of samples taken from bottled water for 48 major water companies in Lebanon showed that samples from 38 companies (79%) were polluted either chemically or biologically or combination of both (CPA 2015).

9.4.3 Unwise Use of Water Resources

Many reasons lead to the misuse of water resources, but the ethical aspect remains the fundamental factor. However, the other factors are also negatively contributing on the consumption of water. Thus, the unwise use of water it is merely wrong behavior that represents a man-made challenge which can be mitigated if consumers take the appropriate actions.

In Lebanon, the intermittent water supply in combination with insufficient amount of water supplied to cope with water demand, results chaotic use of water; and therefore, consumers try use and store water as much as they can, especially that partitioning is an applied method of water supply in Lebanon where water is provided to inhabitants 1–2 days a week.

The largest part of misused water in Lebanon is in agricultural purposes where unorganized irrigation (e.g. furrow irrigation) is widespread. This is almost linked with chaotic abstraction of groundwater and direct pumping from rivers and springs which are based on individual behave. Also, the uncontrolled water exploitation, whether for domestic use or the unaccounted aspects of water makes it aspect of water challenges.

Chapter Highlights
- There are natural and anthropogenic challenges influencing the availability and supply of water resources in Lebanon.
- The major four aspects of natural challenges have been discussed. They include, in a broad sense, the topography, climate, hydrology and the shared water resources.
- Even though, it is not easy to stop these challenges, but mitigation measures can be taken to reduce their impact on water resources and their nexus.
- The author agreed that the impact of human interference on water resources in Lebanon is much more harmful than the natural influencers.
- The principal three aspects of man-made challenges have been discussed. They include, population growth, water quality deterioration and the unwise use of water resources.

References

Bakalowicz M (2009) Assessment and Management of Water Resources with an emphasis on prospects of climate change. In: Policy Dialogue on Integrated Water Resources Management Planning in the Republic of Lebanon. MED EUWI, Beirut

Bou Zeid, E. and El-Fadel. 2002. Climate change and water resources in Lebanon and the Middle East. J Water Resour Plan Manag. 128:5 (343), pp 343–355

CAL (Climatic Atlas of Lebanon) (1973) Meteorological services, vol I. Ministry of Work and Public Transport, 1973, p 31

CAL (Climatic Atlas of Lebanon) (1982) Meteorological services, vol II. Ministry of Work and Public Transport, 1982, p 40

CNRSL (National Council for Scientific Research, Lebanon) (2015) Regional Coordination on Improved Water Resources Management and Capacity Building. Regional project. GEF, WB

CNRS-L/CESBIO (2019) Observation Spatiale de l'Enneigement et Ressources en Eau. Les rendez-vous de l'enseignement supérieur franco-libanais. Beirut, Lebanon February 2019.

CPA (Consumer Protection Association) (2015) Analysis of water quality in bottled water of Lebanon. Unpublished Report, 7 pp

Darwich T, Atallah T, Baydoun S, Jomaa I, Kassem M (2008) Environmental risk of nitrate accumulation in the soil-groundwater system in Central Bekaa Valley, Lebanon

Darwich T, Shaban A, Hamze M (2014) Assessment of climate change impact on water, agriculture and energy in Lebanon. International Conference on: The Water-Food-Energy-Climate Nexus in Global Drylands, Rabat, Morocco, 12–13 June 2014

DGAC (Direction Générale de l'Aviation Civile) (1999) Rapport annuel, Beyrouth, Liban, 32p

Fadel A, Atoui A, Lemaire B, Vinçon-Leite B, Slim K (2014) Dynamics of the toxin cylindrospermopsin and the cyanobacterium Chrysosporum (Aphanizomenon) ovalisporum in a Mediterranean eutrophic reservoir. Toxins (Basel) 6:3041–3057

Fadel A, Atoui A, Lemaire BJ, Vinçon-Leite B, Slim K (2015) Environmental factors associated with phytoplankton succession in a Mediterranean reservoir with a highly fluctuating water level. Environ Monit Assess 187:633. https://doi.org/10.1007/s10661-015-4852-4

Fadel A, Faour G, Slim K (2016) Assessment of the trophic state and Chlorophyll-a concentrations using Landsat OLI in Karaoun reservoir, Lebanon. Leban Sci J 17:130–145. https://doi.org/10.22453/LSJ-017.2.130145

Funk C, Peterson P, Landsfeld M, Pedreros D, Verdin J, Shukla S, Husak G, Rowland J, Harrison L, Hoell A, Michaelsen J (2015) The climate hazards infrared precipitation with stations – a new environmental record for monitoring extremes. Scientific Data. https://doi.org/10.1038/sdata.2015.66

GAIN index summarizes a country's vulnerability to climate change and other global challenges in combination with readiness to improve resilience (2017). http://gain.nd.edu/our-work/country-index/rankings/

Ghaddar N (2003) Climatological data. Monthly bulletin. American University of Beirut, Lebanon

Index-Mundi (2017) Lebanon population growth rate. Available at: https://www.indexmundi.com/lebanon/population_growth_rate.html

IPCC (Intergovernmental Panel for Climate Change) (2010) Working Group I (IPCC-WGI). Climate change: The science of climate change. Cambridge University Press, Cambridge, UK

Khair K, Aker N, Zahrudine K (1992) Hydrogeological units of Lebanon. Hydrogeol J 1(2):34–49. Springer-Verlag

Khair K, Aker N, Haddad F, Jurdi M, Hachach A (1994) The environmental impact of human on groundwater in Lebanon. Water Air Pollut 78:37–49. Kluwer Academic Publications

Khawlie M, Shaban A (1998) Contribution of remote sensing studies to the fractured karstic coastal terrain, Lebanon-Water enhancement or hindrance. Conference on: flow, friction and fracture, AUB Center for Advanced Math. Sciences, Beirut, 1–7/7/1998

Lababidi H, Shatila A, Acra A (1987) The progressive salinization of groundwater in Beirut. Int J Environ Stud 30:203–208

LARI (Lebanese Agricultural Research Institute) (2017) Climatic data. Monthly bulletin. Department of Irrigation and Agro-meteorolog, 16 p

LRA (Litani River Authority) (2017) Rivers discharge records database (Unpublished Report)

MoFA (Ministry of Foreign Affairs) (2018) Climate change profile, Lebanon. Ministry of Foreign Affairs Netherlands. 139p

Nehme N, Haidar C (2018) The physical, and chemical and microbial characteristics of Litani River water. In: The Litani River, Lebanon: an assessment and current challenges. Springer, Cham, 179 p

NOAA (National Oceanographic Data Center) (2013) Lebanon climatological data. Library. Available at: http://docs.lib.noaa.gov/rescue/data_rescue_lebanon.html

PBL (Netherlands Environmental Assessment Agency) (2018) The geography of future water challenges. PBL publication number: 2920. 102 p

Saadeh M (2008) Seawater intrusion in Greater Beirut, Lebanon. In: Climatic changes and water resources in the Middle East and North Africa. Springer, Berlin, pp 361–371

Saadeh M, Wakim E (2017) Deterioration of groundwater in Beirut due to seawater intrusion. J Geosci Environ Prot 2017(5):149–159

Shaban A (2003) Etude de l'hydrogéologie au Liban Occidental: Utilisation de la télédétection. Ph.D. dissertation. Bordeaux 1 Université, 202p

Shaban A (2010) Support of space techniques for groundwater exploration in Lebanon. J Water Resour Prot 5:354–368

Shaban A (2011) Analyzing climatic and hydrologic trends in Lebanon. J Environ Sci Eng 5(3)

Shaban A (2012) A brief review of groundwater resources in coastal Lebanon. In: The INCAM environmental and coastal ecosystem management. CIHEAM, IRD, EU, 213p

Shaban A (2014) Physical and anthropogenic challenges of water resources in Lebanon. J Sci Res Rep 3(3):164–179

Shaban A (2018) Imbalanced water availability/supply in Lebanon. In: Applications imbalance of nature. GEOS Press, Arizona, pp 73–93

Shaban A (2019a) Water resources in Lebanon: challenges in matching availability and supply. COMEST National Seminar. Ethics of Scientific Knowledge and Technology. National Library, Beirut, April 5th, 2019

Shaban A (2019b) Striking challenges on water resources of Lebanon. In: Hydrology: the science of water. InTech Open, London

Shaban A, Hamzé M (2017) Shared water resources of Lebanon. Nova, New York, 150pp

Shaban A, Khawlie M, Abdallah C, Faour G (2005) Geologic controls of submarine groundwater discharge: application of remote sensing to north Lebanon. Environ Geol 47(4):512–522

Shaban A, Faour G, Stephan R, Khater C, Darwich T, Hamzé M (2016) Assessment of coastal wetlands in Lebanon. In: Moran G (ed) Coastal zones: management, assessment and current challenges. Nova Science Publishers, Inc, New York, pp 27–97. isbn:978-1-63485-611-9

Shaban A, De Jong C, Al-Sulaimani Z (2017) New approaches for responsible management of offshore springs in semi-arid regions. Geophysical Research Abstracts. Vol. 19, EGU 2017–19370, 2017EGU General Assembly 2017

SNC (2011) Second National Communication to the UNFCCC. Climate change vulnerability and adaptation. Ministry of Environment & GEF & UNDP, Beirut, 288pp

Telesca L, Lovallo M, Shaban A, Darwich T, Amacha N (2013) Singular spectrum analysis and Fisher-Shanon analysis of spring flow time series: an application to Anjar Spring, Lebanon. Physica A 392:3789–3797

TRMM (Tropical Rainfall Mapping Mission) (2014) Rainfall archives. NASA. http://disc2.nascom.nasa.gov/Giovanni/tovas/TRMM_V6.3B42.2.shtml

USAID (2016) Fact sheet. Climate Change Risk Profile Lebanon. Available at https://www.climatelinks.org/resources/climate-change-risk-profile-lebanon

WB (World Bank) (2017) World development indicators. Available at: http://datatopics.worldbank.org/world-development-indicators/

WEF (World Economic Forum) (2015) Global risks 2015. Insight report, 10th edn. 69p

WHO (World Health Organization) (2006) A Proposal for Updating Lebanese norm of Drinking Water (1999) based on WHO Guidelines

Chapter 10
Proposed Solutions

Abstract "Water in Lebanon is in Jeopardy". Since the beginning of 1990s the water crisis has been occurred in Lebanon and still continuing. Up to date, no improvement in the water sector can be touched. The discharge in rivers and springs has been significantly decreased, it is also the case for many wetlands and lakes. Groundwater is under depletion with sharp decline in the discharge and the abrupt lowering of water table. Quality deterioration became widespread including a surface and groundwater resources where it reached the bottled water. In the view of this unfavorable situation in water sector in Lebanon, questions are always raised: Why there is such a status? What are the reasons behind it? What are the solutions to alleviate the impact of water crisis in Lebanon? What the outcomes of measures and water policies adopted by the government? The most important question remains: What will happen if this deteriorating situation in continues as it is? Actually, there is no define answers for all these questions, notably it is not determined who should answer on them. There must be clear and practical solutions, based on scientific concepts, to face this situation. This chapter presents proposed solutions which are based on the author's expertise and observations. Data and information mentioned in this document were used as a background to build the proposed solutions which represent scientific outlines for further actions.

Keywords Poor supply · Artificial recharge · Strategies · Institutional coordination · Awareness · Water crisis

10.1 Dimensions of Water Crisis

Water crisis is a general understanding with several aspects and dimensions. It is usually linked with themes on vulnerability, resilience, and sustainability. Water crises is, sometimes, named as water security (Varis et al. 2017), disaster-induced water shortage (Kato and Endo 2017) and many other terms such as water scarcity, shortage and water problem. However the term crisis is much more alarming.

A. Shaban, *Water Resources of Lebanon*, World Water Resources 7, https://doi.org/10.1007/978-3-030-48717-1_10

For Lebanon, it is a paradox that water is abundant, as mentioned previously in this documents, while challenges on water supply occur. Thus, all people in Lebanon including different stakeholders and even high-level decision makers, declare that the water sector is totally degraded and moving towards undefined and unfavourable destination.

In this respect, contradictory in determining the status of water sector still exists. While UNUCR (2019) considers that *water sector in Lebanon is working towards its expected outcome by 2020, and consumers are accessing sufficient, safe with reduced health and environmental impacts; nevertheless, the* World Bank (WB 2012) stated that water sector in Lebanon is delivering poor services at a high fiscal and household cost, and water sector inefficiencies (certainly in low collection of tariffs and high water losses) as well as the environmental damages are costing the economy the equivalent of almost 3% of GDP annually.

A socioeconomic survey, done by the author to support the information of this document, has been applied in different regions from Lebanon. A part of the survey inquired about the satisfaction of consumers with respect to water supply, and thus 94% of the answers agreed that water supply and services is poor. This in turn clearly describes water status in Lebanon as "Water Crisis".

However, the impact of water crisis on different sectors, including mainly the nexus of lifestyle-health-agricultures-food-energy; has spatial and temporal dimensions (Table 10.1). These dimensions can be the major index to monitor the changing water status and identifying its optimistic or pessimistic trends. This can be applied once a zonation is elaborated to attribute the crisis dimensions within a define area. Table 10.1, which involves data and information deduced from the resulted analysis (mostly included in this book), summarizes the following zonations:

1. Hydrologic systems: Water crisis may develop within define hydrologic system whether it belongs to surface water bodies (e.g. rivers, lakes, etc.) or to the groundwater reservoirs (i.e. aquiferous rock formations). Therefore, water inputs/outputs in these systems can be analysed.
2. Geographic zones: This is viewed from the land use/land cover basis. It includes the existence of water crisis in particular areas with high water consumption including mainly the arable lands (i.e. cultivated lands) and/or urbanized areas.

10.1.1   Spatial Dimensions

Spatial dimensions of a water crisis imply the geographic distribution of water scarcity and related problems (e.g. pollution), and it can be viewed from the increasing or decreasing trend of the crisis. Thus, the spatial increase in the crisis evidences bad situation on water sector and it points out to pessimistic scenario.

Table 10.1 Spatial and temporal dimensions of water crisis in Lebanon

Zonation		Crisis index		Spatial dimensions	Temporal dimensions
		Volumetric	Quality		
Hydrologic system	Rivers	There is abrupt intermittency in rivers' run-off. Also, the decrease in rivers' discharge exceed 37.5%	Severely contaminated	Run-off over the entire Lebanese territory is totalling a length of about 818 km for the primary watercourses. While, the total area of catchments is 7514 km².	Decreased discharge is about 22 million m³/year. Examples: Increased Nitrite is 4 ppm/year. Increased Total Coliform is 25–40 in 100 ml/year.
	Springs	Large number of springs in Lebanon have been disappeared, and the discharge has been decreased by about 41%.	Moderately polluted	Covering the entire Lebanon where 78% of them located at altitude exceeding 550 m.	Decreased discharge is 13.4 million m³/year.
	Lakes & reservoirs	Obvious reduction in the areas of most lakes and reservoirs (e.g. 26% reduction in the Qaraaoun Reservoir)	Polluted	Estimated area of wetlands in Lebanon is about 42 km² including different scales wetlands.	Estimated areal reduction for all lakes and reservoirs is 10–15%. While, the decreased area is 0.175 km²/year. Examples: Increased Cyanobacteria is 1.5–2 times/year.
	Groundwater reservoirs	The drawdown in the major aquifer exceeded 17%. While, discharge has been decreased by more than 30%.	Shallow aquifers (< 25 m) are totally polluted.	The estimated areal extent of aquiferous rock formations in Lebanon is about 8330 km², where the total thickness of these aquifers is about 1200 m.	The estimated drawdown is about 0.5 m/year. Also, the 30% decrease in water discharge from wells indicates that approximately 1% decrease is annually added. Example: The increased Nitrate in shallow groundwater is 0.5 times per year.
	Snowpack	An increase in the melting duration from 3 to 2 months.	Only dust lamina.	Average area exceeds 1800 km², and lately reported area was more than 6700 km².	Increased melting duration is 3 days/year.
	Wetland	All wetland show reduction in their area. While some of them have been totally dried (i.e. cliffs of Ras Ech-Chekkaa).	Partially polluted	About 15 km² for the large-scale wetlands. In addition to doubled area for the small-scale ones (> 200 wetlands).	Over the last 30 years, one wetland has been vanished and obvious decrease in the saturation rate has been reported.

(continued)

Table 10.1 (continued)

Zonation		Crisis index		Spatial dimensions	Temporal dimensions
		Volumetric	Quality		
Geographic zones	Arable lands	Insufficient irrigation for 30–35% of arable lands	Irrigation using wastewater	Irrigated lands are 1200 km² of Lebanon's area (12%)	12–15 km² of arable lands are suffering from insufficient irrigation every year.
	Urban areas	Water is supplied less than 3 h a day in the major urbanized clusters (WB 2012)	Supplied water is partially polluted. While water from coastal aquifer is polluted	In Lebanon, 87% of population live in urban areas (UN-HABITAT 2008). Urbanized areas occupy 12% of Lebanon's area. Also, the contamination of bottled water has been increased from 36% in 2009 to 79% in 2015.	Domestic water supply is decreasing by about 2 m³/capita/years. Example: the increased contamination in bottled water is about 7% per year.

Table 10.1 shows the spatial distribution of the crisis where the data and information illustrated in the table have been deduced from the resulted analysis (mostly included in this book). Hence, the table shows the following problematic dimensions:

1. The total length of rivers in Lebanon is about 818 km which is covering the entire geography of the country. Hence, the crisis of water decrease and pollution in rivers impacts the basins of these rivers which have a total area of about 7514 km^2 (72% of Lebanon). This means that more than 2/3 of Lebanon is under the impact of rivers' crises.

2. It is similar to the case of rivers, springs are also widespread in Lebanon, even though the largest number of them is located at altitude above 550 m. Therefore, the crisis of water decrease and the moderate pollution level in springs is also influencing the entire Lebanon.

3. The calculated 42 km^2 of lakes and reservoirs are distributed in different Lebanese regions. Thus, pollution and area reduction (& capacity) in these water bodies affect the surrounding regions including agricultural and municipal water. In addition, the regions which are fed from lakes or reservoirs water will be also under the crisis (e.g. Qaraaoun Reservoir and the downstream area under 800 m).

4. The areal extent of the aquiferous rock formations is approximately 1/2 of the Lebanese territory where these formations have different thickness, while the crisis in groundwater is viewed vertically. Hence, the lowering in water table and the declined discharge will affect the 5257 km^2. However, little impact occurs in some localities as a result of the geologic setting of these localities. In addition, the pollution of shallow groundwater occupies mainly the Quaternary aquifer which has an area of about 400–500 m and with depth of less than 25 m.

5. Even though the snowpack area did not remarkably changed, yet the duration of melting has been reduced by 2/3 of the normal rate. This covers the top mountains of Lebanon and specifically areas above 1500 m, where an average area of about 1810 km^2 will be influenced.

6. The spatial dimensions of water decrease in wetlands of Lebanon is not restricted only for the 15 km^2, but the crisis extends, in addition to the surrounding regions, to the habitats of many flora and fauna.

7. The insufficient water for irrigation, as a water supply crisis, occurs mainly in the plains (Bekaa or coastal plain) where it badly affected the crops quality and productivity in more than 140 km^2 among these plains.

8. There are more than 1250 km^2 in Lebanon can be considered as urbanized area where about 87% of population live. Thus more than 3.7 million people in Lebanon are under water crisis including the insufficient supply and contamination.

10.1.2   Temporal Dimensions

The temporal dimensions are represented by the time factor for the changing trends in water crisis. They are also indices for the optimistic or pessimistic conditions. These dimensions are usually expressed by numeric values where a "year" is a time variable selected in this document as mentioned in Table 10.1.

Based on similar data and information listed in spatial dimensions; however, temporal dimensions were calculated as follows:

1. The estimated 37.5% decrease in the discharge of the Lebanese rivers (2800 million m³/year), shows that the total amount of the decreased water over 47 years (1970–2017) was 1050 million m³, which means that about 22 million m³ of water has been decreasing per year, and this is theoretically decreased from the water budget in rivers. This is also accompanied with the increased rate of pollution, such as 4 ppm/year in Nitrite.
2. The total amount of the decreased water in springs is 631 million m³ over 47 years, which is equivalent to about 13.4 million m³ a year.
3. Lakes and reservoirs in Lebanon are witnessing a reduction in their area which was estimated at about 0.175 km²/year. While water pollution is also developed and (for example) the increased Cyanobacteria is increase about 1.5–2 times per year.
4. The crisis in groundwater is vertically acted by the lowering in the water table, which was estimated at 0.5 m/year. While, the pumping is decreasing by 1% every year. While, contamination is also increase, such as that nitrate in shallow groundwater in increased by 0.5 time per year.
5. The speeding duration of snowpack melt was estimated generally by 3 days annually. While, the wetlands are also under water crisis, but no numeric values could determine yet except that these surface water bodies are under destruction.
6. There is increased demand for irrigation water in the arable land. This was estimated at 12–15 km² every year.
7. For the urbanized areas, the domestic water supply is declining by about 2 m³/capita/years. Also, pollution in these areas are well pronounced. A typical example is that the increased contamination in bottled water is about 7% per year.

10.1.3   Future Expectations

It is obvious from the elaborated dimensions of water crisis in Lebanon that there is a movement in the pessimistic direction. Therefore, areas with favourable amounts of water are decreased and joined with abrupt regression in the discharge of water. This is also well observed in the increased annual numerical values and ratio towards water scarcity and developed pollution.

Based on the analysis of water crisis dimensions, there are many possibilities raised, as follows:

1. Rivers:

    Considering the average decrease ration in the discharge of rivers is 37.5% over 47 years, this means that the annual decrease ration is about 0.8%.

    If the status in rivers continues in the same trend with annual decreasing rate which is 22 million m^3; however, the Lebanese rivers will loss more than 200 million m^3 by the next 10 years. Also the continuity of this trend (annual decrease of about 0.8%) points out that the Lebanese rivers will lose their water totally after 125 years.

2. Springs:

    Since the average discharge decrease ration in springs is 41% over 47 years, this means that the annual decrease ration is about 0.9%. Therefore, the continuity of the current trend (annual decrease of about 0.87%) Lebanese rivers will lose their water totally after 115 years.

3. Lakes and reservoirs:

    Generally, the calculated reduction in the area of lakes and reservoirs in Lebanon was 0.175 km^2 per year. This raised an expectation that total area of about 1.75 km^2 will be by 2030. While, the increased pollution will be also developed. For example, the estimated increase in the Cyanobacteria was 1.5–2 times per year, and hence, this will be approximately 17.5 time by 2030.

4. Groundwater:

    Groundwater depletion is well observed in the major aquifers where 0.5 m per year is the average reported drawdown; and therefore, 5 m lowering in the water table is expected to be in 2030.

    If this value is allocated for the areal extent (3645 km^2) of the two major aquifers in Lebanon; therefore, a water volume of about 1.1 billion m^3 will be totally depleted by 2030. While, the regression in the discharge of water from wells, which was estimated at 1% per years, will be reduced at 10% by 2030.

5. Snowpack:

    The rapid melting duration of snowpack will negatively affect the recharge rate and may result floods. Hence, the increased melting duration of 3 day per year, as it was calculated, will be again exacerbated, but this will not be with exponential regime, because the temperature increase (as the main reason behind) will be very changeable and in relatively slow.

6. Wetlands:

    The saturation of surface water bodies in Lebanon are clearly under non-continuity and it is in regression whether in the area or the amount of water they contain. This influences the flora and fauna, as well as the water supply. Even though no numeric measure to point out the degree of regression in wetlands, yet some of them have been vanished and others are under unfavourable conditions. Thus, the continuous in this status is expected to result harmful consequences, within the coming few years, on these water resources.

7. Arable lands:

    The 12–15 km^2 annual increase in water demand in the arable lands will be exponentially exacerbated in the next years, because it is influenced by several factors that make this increase rapid, in particular the dramatic population

increase and demand for water. Therefore, it is anticipated that at least 150 km^2 of the arable lands in Lebanon will suffer from bad irrigation by 2030.
8. Urban area:
It can be said that urban areas in Lebanon are the most influenced clusters in water crisis. This is because water resources in these clusters are few and population size is relatively high. Therefore, the estimated decrease in water quota at 2 m^3/capita per year is alarming, notably that the current water supply is inadequate to cope we water requirements.
Based on the available measurements, it is expected that 20 m^3/capita will be decreased from water quota of the inhabitants in Lebanon by 2030. This is equivalent to about 9% of water needs.

10.2   Proposed Solutions on Surface Water

Availability and loss of surface water resources have been mentioned in several sections of this document. Besides, the urgent demand for water is also mentioned and pessimistic outlooks were almost the most dominant vision for water sector in Lebanon. Therefore, it is normal to harmonize these elements in order to reach the most optimal figure on water resources.

There is still argument about solutions proposed to utilize the available surface water before it is lost. The reason why this has not happened yet can be attributed to either the misunderstanding, limited scientific knowledge, lack of financial resources, corruption or the existing political tension in Lebanon.

Based on several factors, including the expertise and knowledge of the author on surface water resources and their harvesting in Lebanon, the following solutions can be proposed as primary clues for further analysis and then applications. However, the proposed solutions are creditable methods to be developed.

10.2.1   Mountain Reservoirs

Mountain reservoirs are dominant in Lebanon (estimated at 2500 ones) and were discussed in Chap. 6 (Sect. 6.3.2). However, proposing them again in this section is mainly due to their feasibility as an approach of water harvesting, notably where precipitation rate is high enough to be collected.

Not all executed mountain reservoirs in Lebanon are successful, in particular those which have been built in the wrong localities. Therefore, the identification of the suitable sites for mountain reservoirs is significant and must be based on hydrological concepts. In this respect, there are two main sites where mountain reservoirs are usually constructed in Lebanon. These are:

1. Along stream channels: This is almost unfavorable if the bed-load is high and the latter is governed by many factors including mainly the morphometric properties. Thus, why most of mountain reservoirs constructed along stream channels did not succeed and then they usually collapsed/or subjected to intensive erosion and sedimentation.
2. Along snowmelt pathway: It is most feasible and always successful when mountain reservoirs are built along the pathways of snowmelt where slow and uniform run-off occurs. Of course, the characteristic of the terrain surface where the feeding snowpack is located must be investigated.

Therefore, mountain reservoirs are usually found at altitudes above 1500 m where snowpack is almost dominant. However, the following factors must be considered:

1. Terrain characteristics (e.g. slope, channel dimensions, soil type, geology, etc.),
2. Morphometry of the upstream area that feeds the reservoir,
3. Favorable precipitation (i.e. considerable amount of water).

There are three major shapes recommended while constructing mountain reservoirs. These are: (1) rounded, (2) elongated, and (3) triangular.

The dimensions for mountain reservoirs are dependent on the purpose and amount of water required. While, the capacity of these reservoirs in Lebanon usually ranges between 1500 and 5000 m^3.

Shaban and Darwich (2008) introduced a typical design for the mountain reservoirs. This design was successfully applied in many mountainous regions in Lebanon (Fig. 10.1). The advantage of making these reservoirs includes the use of local materials for construction, notably the argillaceous rocks and clays as well as the flattened stones.

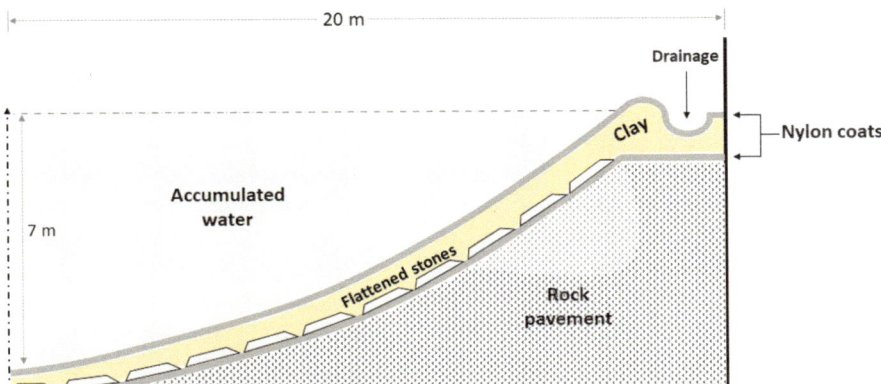

Fig. 10.1 Typical design for mountain reservoirs

10.2.2 Dam Reservoirs

Similar to mountain reservoirs, dams are also known in Lebanon and there are still several dams proposed; however, this aspect of water harvesting is highly recommended in Lebanon even though arguments are raised about their consequence on the environment and water resources. This has been discussed in Chap. 6 (Sect. 6.3.1).

Water in streams and rivers in Lebanon is characterized by rapid run-off along the slopping terrains where the velocity of run-off from the top mountain to the sea may not exceed a couple of hours (i.e. about 5 h). While, the average time between rainfall peak and its appearance into the sea (as plume) was estimated at 2.4 days in all coastal rivers of Lebanon (Shaban et al. 2009).

It is not exaggeration that one of the main solutions for water supply in Lebanon can be solved by executing dams where the resulting reservoirs from these dams are able to capture the water which is now being flow to the sea. While, the arguments about dams location is not a scientifically-based ideas notably that new advanced engineering practices and controls can be applied.

Therefore, dams with different dimensions can be proposed whether on the primary watercourses (& rivers) or along secondary watercourses. While dimensions of these dams are always dependent on the amount of water to be stored.

For the Lebanon's case, dams can be proposed from the dimensional point of view, regardless of the engineering design and properties which are not of the topic in this document. Hence, two major types are suitable as follows:

1. Check dams: these are small-scale dams that can be constructed even on the individual basis. These dams, which can be only few meters in width, are constructed across streams to reduce erosion processes after lowering water flow energy, but they collect water and give enough time for infiltration into substratum. They also work in improving the moisture of bed sediment in adjoining areas (Hassanli and Beecham 2009).

   Due to the narrowness of valleys in Lebanon and the steep sloping channels, check dams can be constructed along the 3rd stream order and higher (i.e. 2 and 1), where the bifurcation ratio is suitable for water accumulation.

   The topography of Lebanon enables constructing frequent check dams along the same stream, which can be described as "ladder" check dams (Fig. 10.2). This was adopted by Nyssen et al. (2004) who agree that the effective check dams often found with a spillway, an apron, a concave plan form (looking down slope) and they were are constructed at vertical intervals and with heights that result in a negative gradient for the line connecting the spillway and the base of the upstream check dam (Fig. 10.2).

   Therefore, several hundreds of check dams can be proposed within each watershed where water capacity of at least 1500 m^3 can be attained for each.

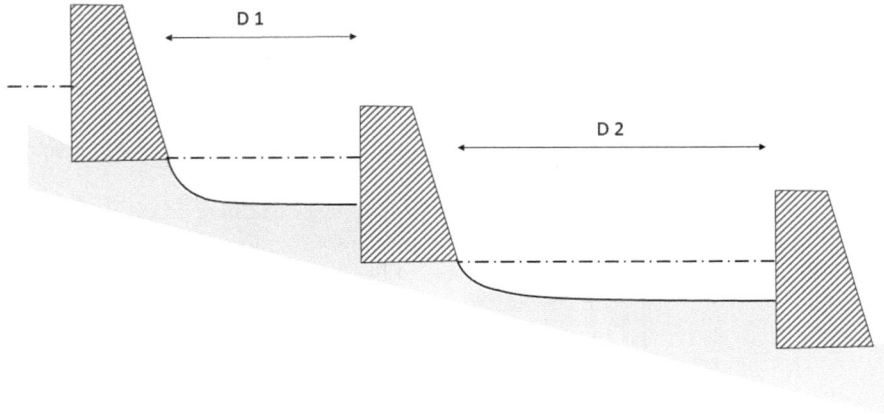

Fig. 10.2 Typical example check dams with ladder configuration. (Adapted from Nyssen et al. 2004)

2. Earth-fill dams:

These are describing the relatively large-scale dams which are executed on the governmental level. Such dams are found in Lebanon, but they are still inadequate with respect to the volume of water that annually lost to the sea without any benefit.

There are many classifications of Earth-fill dams where these classifications are based on:

– Materials used (e.g. rigid, non-rigid dams),
– Function (e.g. detention, storage, diversion dams),
– Hydraulic design (over flow, non over flow dams)
– Stability conditions (arc, gravity, cantilever dams).

These dams are expected to be at least 100 m length and exceed 10 m height. They are often constructed of compacted earth where a zone in the middle (core) made of low permeability material, a permeable part growing gradually outward called a filter on the two sides covering the core which is usually composed of argillaceous or impermeable rocks materials to prevent water passing through the dam.

There are 18 large-scale dams proposed by MoEW in Lebanon within the horizon of 2030 where the capacity of these dams ranges between 4 and 128 million m^3 and if they are constructed, they are anticipated to store an annual volume of 1100 million m^3 (SNC 2011).

Nevertheless, the geomorphological characteristics of Lebanon in combination with its geology and hydrology makes it reasonable to propose at least doubled number of the proposed ones by MoEW.

10.2.3 Water-Convey Canals

Surface water convey canals can be observed in Lebanon, but if their utility is accounted, therefore, the existing canals are still few. The largest part of these canals occur in the coastal plain, such as those executed by LRA near Sour area in South Lebanon.

However, the concept of water convey canals proposed here is different than those constructed in the coastal plain. This is because the proposed canals is to be established in the upstream regions (i.e. almost the mountainous regions). They can be delivered from the primary watercourses (rivers) or the major intermittent streams.

The typical cross-section dimension of these canals is about 2 m². The purpose of constructing these canals is to divert river or stream water, which can be connected with ground ponds at a distance, from the major watercourse and in the proximity of locality where water will be consumed, e.g. cultivated lands, urban areas, etc. (Fig. 10.3).

Normally, the morphometry of the region, where the convey canals are proposed, must be well assessed. While, the locality of joining the canals with watercourse is most applicable at stream meanders, diversions and confluence.

10.2.4 Rooftop Rainwater Harvesting

Rooftop harvesting (RTH) has recently become cheaper and more feasible in performance. It is usually used to capture rain water all winter long and then to provide with considerable amount of water during the dry season. Interest in RTH technology is reflected in the water policies of many developing countries, where it is now cited as an alternative source for household water.

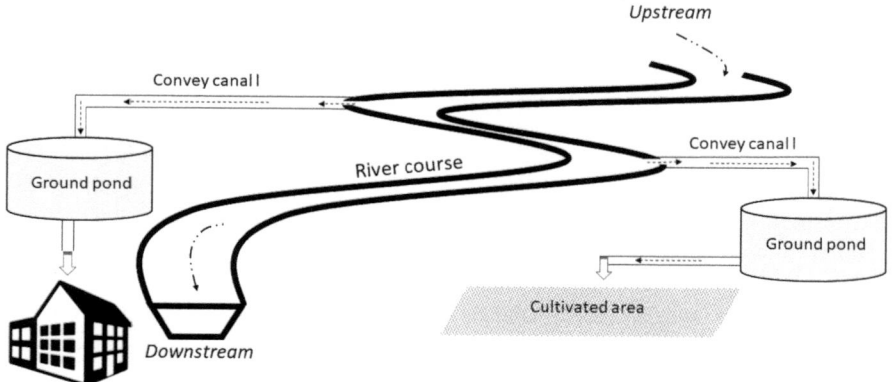

Fig. 10.3 Typical convey canal along the Litani River near Sour area

The idea behind RTH is viewed from the fact that this technique is not influenced by the problem of rainfall patterns which has been changed lately into torrential; therefore, water can be captures on the rooftops area and then stored for the time when water supply becomes slow (e.g. summer season).

Moreover, the increased urban settlements, which cover part of the land surface, can be utilized to execute RTH. Hence, it can be a solution for the decreased recharge rate due to existence of these settlements. In addition, RTH can provide mostly pure water and reduces the probability of floods and chaotic overland flow of water.

Simply, RTH technique is the catchment of rainwater in either on inclined roof covered by roof-tiles or on flat roofs. For the inclined roof with roof-tiles a large pot/vessel kept beneath the edge of the roof. While for flat roofs, water is directly collected on the roof. In both cases, the collected water from is diverted into drain pipes/gutters fixed to roof edge along which water is flow down-ward to tanks or ground ponds made of iron sheets, cement or bricks.

A study on RTH has been carried out and supervised by the author. It has been applied to pilot areas from Lebanon including rural and urban areas with different physiographic characteristics. The study showed that RTH can contribute by approximately 75–80 million m^3/year (Kashoua 2015). This is equivalent to about 18 m^3/capita/year.

10.2.5 Snowpack Reservation

Mountain crests of Lebanon and the adjacent slopes (almost above 1500 m) are known by frequent snow cover (Shaban et al. 2014), such as in mountains of: Hermoun, Niha, El-Barook, Al-Kneiseh, Sannine, El-Mezar, El-Mnaeita, Al-Makmal and El-Kamouaa (Fig. 10.4). This snow cover is mainly accumulated on the exposed rocks of the Cenomanian and Kimmeridjian formations which are both characterized by dense fracture systems and karstification. Therefore, these two rock formations represent potential recharge zones and excellent aquifers and then forming the major groundwater reservoirs of Lebanon.

About 2350 km^2 of the Cenomanian and Kimmeridjian rock masses are located above 1500 ma altitude. This area occupies more than 1.7×106 million m^3 as SWE, and it may exceed 2.4×10^6 million m^3 according to Mhawej et al. (2014). This motivates conserving the snowpack on mountain crests in Lebanon order to remain this water resource uninfluenced.

However, the areas where snowpack is accumulated on these mountain have been invaded by several unfavorable activities done by human (Shaban and Darwich 2011); therefore, these elevated geographic areas should be adopted as natural reserves likewise to other natural reserves located in Lebanon. Thus, the unfavorable activities on mountain crests mainly include the following:

1. Excavating snowpack and put it in the existing mountain reservoirs, and consequently the snowmelt can fill these reservoirs,

Fig. 10.4 Snowpack on Jabal Sannine of the Cenomanian rocks

2. Increasing livestock farming which results contamination reflected on water
   from snowmelt,
3. Increased uncontrolled touristic activities where snowpack is destructed and then
   misbalancing the hydrologic system.

10.3   Proposed Solutions on Groundwater

In Lebanon, usually solutions on water resources are oriented towards surface water
resources while groundwater has little attention except that artificial recharge is
mentioned in some studies (Masciopinto 2013; MoEW and UNDP 2014). Hence,
the proposed solutions here are not necessarily taken from the exploitation point of
view but in the enhancing the recharge rate as well.

10.3.1   Artificial Groundwater Recharge

Artificial Groundwater Recharge (AGR) has become a successful technology
applied in water resources management and for the adaptation measures to the
changing climatic conditions. This technology can utilize the excessive surface

water that would otherwise be lost into the sea like the case in Lebanon where carbonate rocks extend along the coastal zone. This zone occupies more than 70% of the total population and water is dramatically demanded.

AGR is commonly applied in the Arab Region where there is similarity with geologic setting of the coastal zone of Lebanon. Therefore, they were applied either as large productive projects or to experimental projects (UNEP-IETC 2001).

1. The problematic:

As a matter of fact, the coastal zone in Lebanon is witnessing excessive groundwater abstraction due to the increased demand. This in turn resulted exacerbation in water depth even in the major carbonate aquifers where water table has been lowered to more than 25 m. As a consequence of this hydrologic phenomenon, saltwater intrusion has been spread even several kilometres on-land. In addition, the rapid surface water flow in rivers and streams as well as the overland flow on terrain surfaces, makes it necessary to reduce and capture the amount of water lost into the sea.

Therefore, the application of AGR in the coastal aquifers of Lebanon will contribute in:

– Stabilizing water table depth,
– Purification of water quality,
– Avoiding/reducing saltwater intrusions,
– Mitigating floods and erosion.

2. Geo-hydrology of the coastal zone:

95% of the coastal zone of Lebanon is occupied by carbonate rocks and the related belonging rocks where the Neogene, Eocene, Cenomanian rock formations are dominant with some portions of Senonian marl and proximate Jurassic limestone. Therefore, limestone and dolomitic limestone constitutes about 82% of the coastal zone. The distribution of these lithologies is interfered by several rock structures including faults, folds with abundant fracture systems and karistified rock masses.

As a predominant structural feature in the coastal zone of Lebanon, the titling of bedding plans seaward is always observed where the dip angle ranges between 8 and 24 degrees and the rock masses are almost trending in the E-W direction.

From the hydrgeologic point of view, the existence of intervening of permeable and porous rock masses with impermeable layers (i.e. marl and argillaceous materials) makes it suitable for groundwater storage. Nevertheless, the remarkable inclination of the bedding plans as well as the dominant faults and other rock deformations, enhance water seeps seaward. A typical geologic cross-section is illustrated in Fig. 10.5 where the inclination of rock bedding plans and the interbedding of diverse permeability rock layer are illustrated.

Another remarkable hydrgeologic phenomenon occurs along the Lebanese coast where the marl of the Senonian rock formation is overlying the Cenomanian limestone and the contact is almost extending along the shoreline; therefore, the marl rocks create a hydrgeologic barrier that isolates the saltwater from freshwater

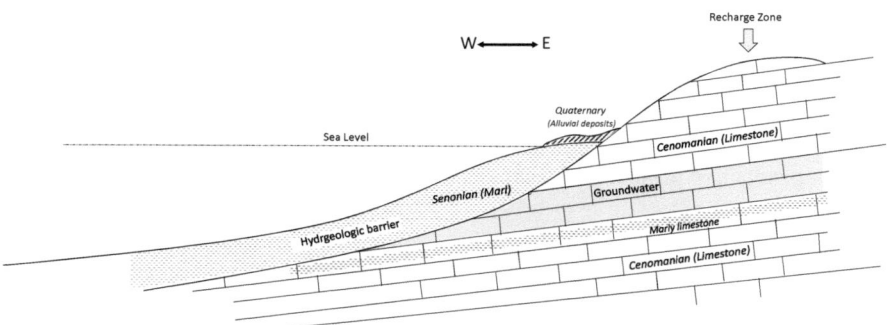

Fig. 10.5 Schematic cross-section, typical for hydrogeology of the coastal zone in Lebanon

interface (Fig. 10.5). This phenomenon works on the coastal zone of Lebanon at different levels; hence, it sometimes completely separates freshwater from saltwater and then freshwater sometimes is found in boreholes even when they are dug very close to the shoreline.

3. AGR technology:

There are several methods applied for AG where all aiming at feeding surface water into groundwater aquifers, notably in the coastal zones. All these methods follow two major concepts:

– Infiltration technology: This technology is applied to promote the infiltration rate from terrain surfaces where aquiferous rock formation are extending under these surfaces. Therefore, two major methods are always followed. These are:

   (a) Recharge dams, where dams along streams are constructed to retard run-off water and infiltrates it downward. This has been successfully applied in many regions. For example, re-charge dams were built along Najran Valley in Saudi Arabia and Al-Khoud valley in Oman where they capture 83 and 11.5 million m³/year; respectively, and then the largest portion of this water has been infiltrated to substratum. (UNEP-IETC 2001).

   (b) Recharge basins, where pits are excavated and gravel are put on the top of these basins to facilitate water percolation. A good example has been reported by Al Saud (2015) who proposed "Infiltration Domains". These domains are artificial pits along streams and low lands where they can be filled by rock aggregates and debris wherein water can percolate rapidly and recharges the underlying rocks.

– Injection technology: This is done by direct pumping, through boreholes, of surface water into the underlying stratum in order to replenish the beneath aquiferous rocks. This has a wide range applications. For example, in Qatar this has been applied in Rawdat Al Faras and Al Hashem where 30% in groundwater recharge was reported (UNEP-IETC 2001). Thus, the following technical aspects can be applied.

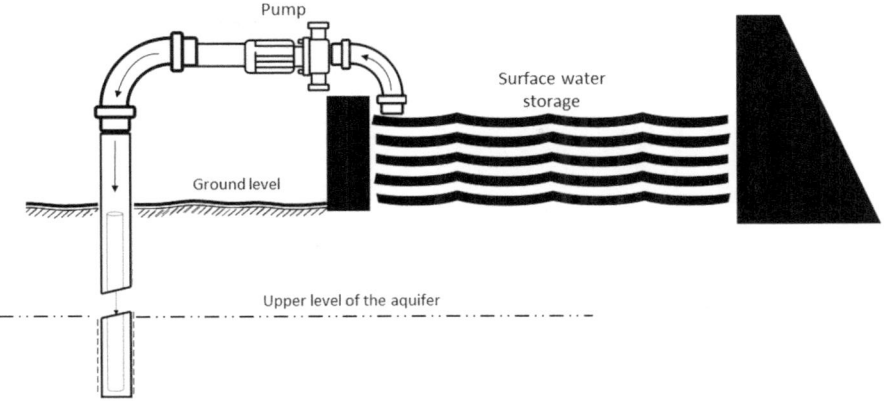

Fig. 10.6 Technology of surface water injection into underlying stratums

(a) Direct injection, where water can be directly pumped from adjacent river water and injected into boreholes. Water can also pumped from flooded areas or along streams after retarding water flow by constructing small dams, pools or any terrain feature that can collect water. Then water is injected through pipes into substratum at different depths depending on the depth of aquiferous rocks (Fig. 10.6).

(b) Spontaneous injection, where boreholes are dug in sites where water may accumulate/or along the pathways of surface water. Therefore, screens are fixed on the tops of these borehole. These screens must have considerable height (about 1.5 m) and should filled with graded gravel to give a chance for the accumulated/running water to enter through them. Also, the terrain surface around the borehole must filled with soil, alluviums, gravel and rock debris to enhance the infiltration rate. Thus, when water reaches the site and its proximity, it will flow down into through the dug boreholes where aquiferous rocks are found beneath (Fig. 10.7).

4. Controls of AGR technology:

There are several controlling factors and measures should be considered while proposing AGR. Any erroneous work in these controls will result unsuccessful application of AGR. Also, these controls, on many instances, differ from one region to another. They can be summarized as follows:

– The geology, geomorphology and hydrology of the area of concern must be well understood, with a special emphasis on drainage morphometry and lithological characteristics.
– Volume of water that may be accumulated on terrain surface or water fluxes along streams should be calculated,
– Many hydrologic properties should be primarily identified in particular the porosity, permeability and hydraulic conductivity and specific yield,

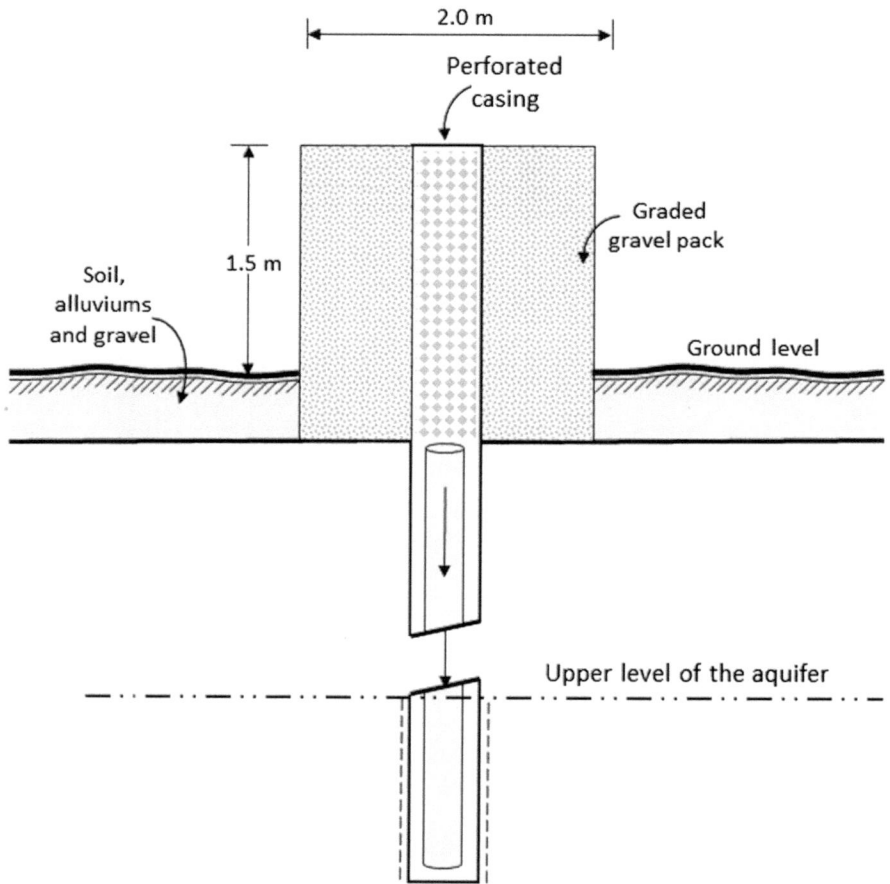

Fig. 10.7 Spontaneous recharge by surface water into underlying stratums

- The depth of upper and lower levels of aquiferous rock formation must be determined,
- The dominant inclination (i.e. dip) of bedding plans should be measured in order to trace groundwater flow direction,
- Rock deformations, notably faults and fractures must be recognized, and they can be favourable site for spontaneous recharge,
- Sub-surface karistified rock masses should be avoided as much as possible, especially if the karstification is not delineated,
- Diameter and screen types of borehole for surface water injection must be selected according to the identified hydrologic properties and the depth of aquiferous rock formations,
- If water wells are present in the area of concern; therefore, their number, depth to water, and discharge from these wells must be identified.

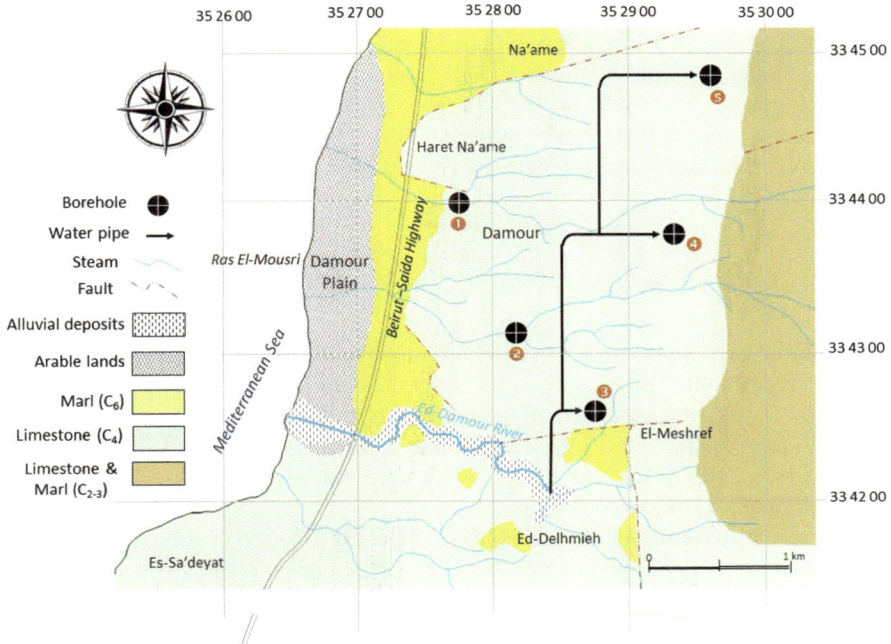

Fig. 10.8 Typical example for selecting localities for groundwater recharge

5. Proposed AGR sites:

A hydrgeologic assessment, integrated with the mentioned controls for AGR, must be perquisite before any carrying out the work. Moreover, the underlying sites for the depleted aquifers should be determined and this can be well recognized form the abstracted groundwater including its dis-charge rate and regime as well as the quality of abstracted water. This is, as mentioned before, attributed to the stress on the coastal aquifers which are considered as major source of groundwater for dense populated urban settlements.

In this regard, the proposition of AGR sites is preliminary, and it needs in-depth assessment when the application of AGR is adopted. As an example, a pilot area (approximately 35 km^2) has been illustrated in this section. Hence, the proposition here is just to introduce the concept of AGR as an optimal solution for groundwater management in Lebanon.

For the selected sites, as shown in Fig. 10.8, surface water sources whether static/ or dynamic were considered, as well as the positioning of the coastal aquiferous rock masses was identified. Thus, Fig. 10.8 reveals a typical example of selecting localities for surface water injection into the underlying rocks. It shows that the injection can be in the coastal aquifer of the Cenomanian rocks where saltwater intrusion and declined water table and discharge occur. It is clear in the figure that boreholes number 1 and 2 are located on streams conjunction where considerable amount of water can be collected. While, boreholes 3, 4 and 5 are receiving water from the Ed-Damour River where pipes can be connected to convey water from the river to another localities as shown in Fig. 10.8.

In this respect and build on the natural hydrgeologic setting and the controls mentioned in point 4; there can be at least 60 boreholes proposed for surface water recharge into the beneath aquifers in the coastal zone of Lebanon.

10.3.2   Capturing Groundwater Discharge into the Sea

Large amounts of groundwater in Lebanon flows to the sea, as sub-marine springs, without any benefit. This was discussed in details in Chap. 4, Sect. 4.3. According to Shaban (2003), the estimated volume of groundwater from these springs is about 410 million m^3/year, which is almost equal to the run-off in three coastal Lebanese rivers. Hence, a part of this non-conventional water resource must be utilized instead of flowing without any exploitation.

Few studies have been done in Lebanon on the exploitation of submarine springs and all were not encouraging towards the capability of invest this water resource. One of these studies which was obtained by Ayoub et al. (2002) declared that the exploitation of the submarine springs could face many problems as it is technically difficult, financially expensive, and yields a qualitatively unacceptable water.

In many coastal regions worldwide, where there is similar geological setting like Lebanon, there are many attempts to tap groundwater flowing into the sea, but these attempts were not satisfactory from to the technical point of view (Bakken et al. 2012; Xuan and Holly 2019).

Therefore, the most feasible approach to tap groundwater before it flows into the sea, as sub-marine springs, is to determine its sources on-land. In other words, the aquiferous where sub-marine springs are fed from must be identified and then exploited.

In this respect, Shaban et al. (2005) put the preliminary bases for the methodology upon which groundwater flows to the see can be tapped. This was based identifying three major hydrogeological component as follows:

- Recharge zone (on-land), where surface water percolate in the beneath rock masses,
- Location of the sub-marine spring whether directly on the shoreline or at a range off-shore,
- Recognizing the geologic controls which are represented by hydrologic routes that connect between the above two components. These routes can be faults, fissures and karstic conduits.
- In order to elaborate these three components, a number of tools were utilized including mainly, the geological and hydrogeological maps, thermal satellite images (e.g. Landsat, Aster, etc.), and this was integrated with field measures. Therefore, the identified terrestrial feeding sources of sub-marine springs along the Lebanese coast were identified as shown in Table 10.2.

Table 10.2 Major geologic controls and terrestrial feeding sources of the sub-marine springs along the coastal zone of Lebanon

#	Sub-marine spring locality	Geologic controls	Source areas on-land[a]	Rock formation
1	Minieh	Alluvial deposits on carbonate rock	Jabal Terbol, 9 km	Cenomanian
2	Hai el-Maqateh	Ditto	Ditto	
3	Bahsas	Karstic conduits	Ras Masqa, 4 km	Miocene
4	Bahsas –Abu Halqa	Karstic galleries and fractured rocks	Ditto	
5	Chekka-1	Karstic galleries	Btourram, 10 km (Miocene) & Mezyara, 18 km; Deir Billa, 13 km)	Cenomanian-Senonian
6	Chekka-2			
7	Chekka-3			
8	Chekka-4			
9	Fadaous	Karstic galleries and fractured rocks	Kour, 7 km & Assia, 12 km	Cenomanian
10	Madfoun-1	Fault alignment	Bechaleh, 17 km &Toula, 8 km	
11	Madfoun-2	Ditto	Chikhan, 3 km	Turonian
12	Wata el Borj	Ditto	Kfer Hatta, 6 km	Cenomanian
13	Helweh- Mar Jerjes-1	Faults and fracture systems	Aidamoun, 1.5 km	Turonian
14	Helweh- Mar Jerjes-2	Fault alignment	Daher Hassan, 4.5 km	Cenomanian
15	Helweh- Mar Jerjes-3	Local fracture systems	Local Cenomanian rocks	
16	Tarol-1	Fault alignment	Jabal El-Ghorab, 3 km	Cenomanian
			Jabal Tartij, 25 km	Jurassic
17	Tarol-2	Tilting of bedding plans	Littoral rock masses	Cenomanian
18	Halat	Fault alignment	Jabal El-Ghorab, 5 km	Jurassic
19	Bouar-1	Karstic galleries	Fatqa, 4–5 km	Cenomanian
20	Bouar-2			
21	Tabarja	Fault alignment	Defane, 2 km	

(continued)

Table 10.2 (continued)

#	Sub-marine spring locality	Geologic controls	Source areas on-land[a]	Rock formation
22	Ma'ameltein-1 el-Borj	Ditto	Jabal Moussa, 11 km & Jabal Tartij, 27 km	Jurassic
23	Ma'ameltein-2 el-Mahatta	Karstic conduits	Dlebta-Harissa, 2 km	
24	Ma'ameltein-3 el-Mahatta			
25	Jounieh port-1		Aintoura, 2 km	
26	Jounieh port-2			
27	Dbayeh	Fault alignment	Beit Chabab, 10 km & Qornet El-Hamra, 22 km	
28	Khaldeh	Fault alignment	Souq El-Ghareb-Chouifat, 8–10 km	Aptian-Albian-Cenomanian-
29	Doha1	Tilting of bedding plans and fracture systems	Mejedlia-Aramoun, 15–18 km	Cenomanian
30	Doha 2			
31	Damour-Saadyat 1	Karstic galleries	Debieh, Borjaee, 20–22 km	Cenomanian
32	Damour-Saadyat 2			
33	Damour-Saadyat 3			
34	Damour-Saadyat 4	Tilting of bedding plans	Barja, Chehim, 5–7 km	
35	Saadyat 1			
36	Saadyat 2	Fault alignment	Majdlouns, El-Magharieh, Chehim, 14–16 km	
37	Saadyat 3			
38	Oudi Ez-Ziena			
39	Er-Rmayleh 1	Fault alignment	Marous El-Fawqa, 10–12 km	Cenomanian-Senonian
40	Er-Rmayleh 2			
41	Sarafand-Aqbieh	Fracture systems	Tefahta, Baysarieh, 6–8 km	Eocene
42	Sarafand 1	Fracture systems, rocks tilting	Tefahta, Bablieh, 5–7 km	
43	Sarafand 2			
44	Khayzaran	Karstic conduits	Qa'akaiet Es-Snouber, 5–6 km	Cenomanian, Eocene

45	Saksakieh	Karstic conduits/fault	Nsarieh, 1–2 km	Eocene
46	Loubia	Tilting of bedding plans		Eocene
47	Adloun-Nsarieh	Faults/fracture systems	Bablieh, Khartoum, 7–9 km	Cenomanian, Eocene
48	Adloun	Fracture systems	Daher Mchar-Daher Katrif, 3–4 km	Eocene
49	Ras Mienet Abou-Zaid	Fault alignment	En-Namirieh, 18–20 km	Cenomanian
50	Abou Al-Aswad	Alluvial deposits	Flood plain of Litani River, 1 km	Quaternary
51	Ras Mienet Chaourane	Fault alignment/karstic conduit	Douiar, Insar, 15–17 km	Cenomanian
52	Boroghlieh			
53	Boroghlieh	Fault alignment	Zrarieh, 7–8 km	
54	Boroghlieh off-shore	Fault alignment/karstic conduit	En-Namirieh, Douiar, 18–20 km	

[a]The mentioned name of a village is a local site representing a larger zone

10.4 Mitigation of Water Pollution

It can be said that the problem in water sector in Lebanon is equally divided between the deficit in water supply and the pollution of water resources where the latter can be much controlled if proper management is followed by different stakeholders. Therefore, mitigating water pollution is the responsibility of individuals, institutions and decision makers at the same level.

The percentage of naturally contaminated water in Lebanon does not exceed 10% where the physical contamination is almost occur as turbidity. While, the negative interference of individuals is largely contributing in water pollution starting from water for household uses.

Besides, looking to the problem of water pollution from the institutional level, thus it is estimated that about 310 million m^3 (92% of Lebanon's sewage) of water is annually flowing to the sea without any treatment (W.B. 2012).

There is real threat in Lebanon from water pollution whether on human health and even on his life, and then rapid solutions should be employed in parallel with environmental legislations and controls as well as following ethical manner for water use. For this reason, the time factor in resolving the problem must be considered. Therefore, in a simple matrix analysis, the aspects of water pollution can be approached by the proposed solution whereas the temporal dimensions for actions should be described as shown in Table 10.3.

Table 10.3 Aspects of water pollution, proposed solutions and their time factor

Aspect of water pollution	Proposed solution		Temporal dimensions
Supplied water for domestic use	Periodical quality control Executing/maintaining sanitation systems	Environmental legislations and controls Following ethical manner in water use	Immediate
Polluted irrigation water	Securing sufficient water for irrigation Adoption of drought-tolerant crops		Short-term[a]
Rivers' water pollution	Executing wastewater utilities Institutional constraints should be reduced Establishing sanitation systems in the riparian zones		Moderate-term[b]
Contaminated groundwater	Controlling groundwater abstraction Reserving surface water recharge zones Adopting waste disposal management approaches		Precautionary to moderate-term
Saltwater intrusion	Avoid excessive pumping from of coastal aquifers		Precautionary
Pollution of Qaraaoun reservoir	Urban planning must be executed, with special emphasis on developing infrastructure and waste disposal management plans.		Moderate-term

[a]Short-term: is to be implemented within less than 1 year
[b]Moderate-term: is to be implemented within less than 2 years

10.5   Economic Policies

Economics polices play a significant role in water management where they can be a powerful analytical tool for water allocation and decision making, as well as they support the establishment of policy instruments (Berbel et al. 2017). Hence, the existed water policies are old and they must consider the significant changes happened and also those changes which may happen in the future. Hence, economic policies must address diversifying social interests that are likely to be involved in chaos, notably in using the resources such as water resources.

In Lebanon, actions to secure and enhance water sector become a priority. This has been lately given attention by local and foreign entities, including public and private institutions which was supported by national and international entities (e.g. UNDP, USAID, IDRC, etc.) and foreign donors in combination with governmental sector.

The actions which were taken primarily by international entities and secondarily the regional ones and donors, have the most impact in improving the water sector in Lebanon. This is merely attributed to the financial resources introduced by these entities. While, the national actions are almost with minimal impact if compared with the former two ones (Shaban 2016).

So far, it cannot be considered that there are water policies implemented in Lebanon. There are only actions and plans taken by national and international institutions to secure adequate and pure water. The actions taken by the national institutions can be summarized as follows:

- Adoption of Law 221: Executed by MoPW (2000) which is a legal framework to set measures for optimal water and wastewater facilities (ESCWA 2010),
- National Water Sector Strategy-NWSS: In traduced by MoPW (2010), and it was established with the participation of national stakeholders and international donors,
- Creating water establishments (WEs) under the mandate of MoPW (2013). This includes five establishments assigned by regions in coordination with the MoPW in order to govern water supply and monitoring,
- The MoPW lunched a number of capital water projects on the national level. This includes mainly the execution of large-scale dams, supply systems, groundwater extraction and treatment planets (Shaban 2016).

On the international basis in combination with the Lebanon's government, there are many actions done whether as executive projects or as studies. Major examples can be illustrated as follows:

- Lebanon's 1st National Communication on Climate change obtained by UNDP and Ministry of Environment-MoE (FNC 2009),
- Environmental fund for Lebanon by the German Ministry for Economic Cooperation and Development –BMZ in 2006,
- Climate change mitigation project by Italian Cooperation & MoE in 2011,

- Lebanon's 2nd National Communication on Climate change vulnerability and adaptation obtained by UNDP and MoE (SNC 2011),
- Improved Water Management for Sustainable Mountain Agriculture by ICARDA in 2012,
- Lebanon's 3rd National Communication to UNFCCC, obtained by UNDP and MoE, 2015 (TNC 2015).

Despite the actions taken at all levels to improve the situation of the water sector in Lebanon, yet there has not been any enhancement reported in this sector so far. Therefore, new economic policies, as a supporting instrument for enhancing the water sector, became a must.

According to Shaban (2016), the cost for implementing water supply in Lebanon for the consumers exceeds several times the refunds paid by consumers. This can be a result of the following:

1. Lack of effective legislations and laws to control water consumption,
2. No water metering is applied in many Lebanese regions,
3. Partitioning of water supply created chaos and motivated individuals not to paying for water,
4. Poor maintenance of water supply systems, as well as the increased quality deterioration,
5. Self-dependant of consumers to secure water needs, such as rooftop harvesting, mountain lakes, water storage, etc.
6. The increased water trading makes it difficult for consumers to pay for additional costs to the cost of water received from the formal water sector,
7. The existing political situation and the lack of water governance to set the effective and appropriate regulations and systematic approaches for water supply.

Based on the unfavourable situation in water sector in Lebanon, economic policies can be proposed as creditable solution where they can be best suited to foster an efficient allocation and use of water, reduce harmful exposure and impacts on the communities and environment, and protect natural capital (Shaban 2016). These policies can be under two major pillars: the cost paying and adopting policies (Fig. 10.9). Therefore, the following are proposed items for applying the successful economic policies in Lebanon:

1. Cost paying:

   - Adopting water tariff:

Tariff represents partial reimbursement of water supply cost where it is usually less than the production cost. This is well known in regions where governments subsidize the difference between production cost and the adopted tariff. Nevertheless, tariff is partially applied in Lebanon where it is found (if exists) with two aspects:

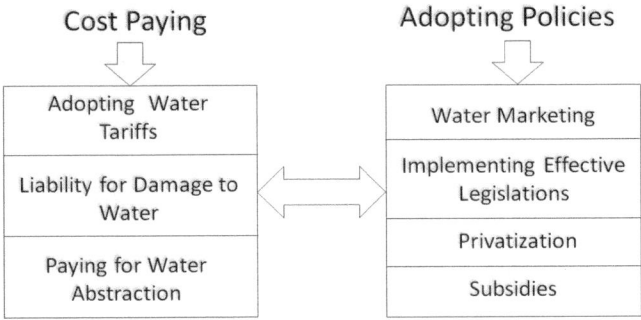

Fig. 10.9 Proposed economic policies for Lebanon

(a) Tariff according to the consumption rate,
(b) Tariff paid as sum lump for annual consumption.

Therefore, adopting tariff becomes necessary in Lebanon to overcome the financial shortage in water sector. In order to activate this measure; however, it must be imposed with legal approaches and more certainly by regular water pricing which can be achieved if water metering (fixing gauges) is applied in all Lebanese regions.

Adopting tariff regulates water consumption and adding financial resources to support the operational and maintenance implements for water supply, it also motivates consumers to use water efficiently.

– Liability for Damage to Water:

Environmental liability systems are applied to internalize and recover the costs of any damage caused by people, and then to make them responsible on the negative consequences, such as damaging networks and contaminating water resources and then they must pay for their behaviour (Shaban 2016). Thus, to that extent environmental liability laws are a fundamental expression of the polluter-pays principle (Kraemer et al. 2003).

Damaging the networks (e.g. pipelines, infrastructure, etc.) of water resources and storage sites (e.g. lakes, ponds, etc.), as well as polluting surface and groundwater are common in Lebanon, especially in the absence of governmental controls and the increased informal human settlements. This needs executing rapid implementations for controlling different localities, notably where human activities are chaotically expended (e.g. along the Litani River watercourses). This can be supported by designating environmental policemen especially for monitoring unfavourable behaves (Hamzé and Shaban 2018).

Liability for damage to waters, as a punishment for individuals and institutions caused water pollution and sources damaging, will reduce pollution rates, and thus securing safe water sources plus saving money for water purification.

– Paying for water abstraction:

Usually, there is tax exemptions or reductions for the abstraction of small quantities of water which is often applied for farmers or some industries as incentives to promote such activities. However, this encouraging regime has been disrupted in Lebanon as a result of the increased demand for water and the chaotic water abstraction.

Controlling water abstraction requires monitoring system to record the abstraction rate from different water resources, specifically from boreholes, rivers and springs. Hence, there is a significant need to meter water abstraction and then paying taxes accordingly. It is in turn will reduce the unwise-use of water resources and also contributes by additional financial resources for the water sector.

2. Adopting policies:

– Water marketing:

This is a widespread task implemented by the private sector where informal water trading is applied and it does not support the formal water sector. Thus, water trading is often characterized by chaotic regime where no fixed or regulated price of water is followed. Hence, water price is determined according to water demand and the availability from sources, whereas, these sources are under the mandate of the formal water sector.

In order to avoid the chaos behavior in exploiting water in uncontrolled trading, there is a great need to propose establishing water s stations and related supply networks. These stations, with controlled quality, must be with diverse dimensions according to the geographic localities, population size and water demand (Shaban 2016). Thus, water marketing can follow financial interest, but within the limited cost. This may result widespread water supply sources with acceptable price and safe water quality.

– Implementing Effective Legislations:

Legal instruments play a major role in that they relocate the proposed economic policies from proposition to execution. Legal instruments; therefore, compose effective instrument to reform and adjust public actions. For this reason, new legislations should be proposed to regulate the existed water issues on the individual and institutional levels.

In Lebanon, adopting updated water legislations from virtual laws into effective actions will be a key solution for several unorganized works taken whether by individuals or institutions and these legislations represent a task force for the economic aspects.

– Privatization:

Even though, it is always rejected by many stakeholders, yet privatization extends chance to the private sector in the provision of water services and sanitation. Therefore, privatization can be an instrument reducing the impact of the overload on water sector, notably in a country like Lebanon where financial resources are not available to cope with needs for water production. Hence, private sector participation leads to improvements in the efficiency and service quality of utilities.

According to Shaban (2016), the implementation of privatization can be either one of the following: (1) Full privatization where assets are permanently sold to a private investor and (2) Public-private partnership (PPPs), ownership of assets remains public and only specific functions are delegated to a private company for a define time period.

– Subsidies:

The Organization for Economic Cooperation and Development (OECD) defines subsidies as government interventions by either direct (e.g. cash grants, interest-free loans, etc.) or indirect (tax breaks, insurance, low-interest loans, rent rebates, etc.) payments, price regulations and protective measures to support actions that favour environmentally-unfriendly choices over environmentally-friendly ones.

Thus, subsidies are instituted either to compensate users for a cost they incur in response to a required works or a prohibition, or they are help to set the necessary incentives for achieving a desired works.

In Lebanon, where financial resources are few enough to ensure subsidies, these resources can be available through the implementation of proposed economic policies. In addition, the adoption of subsidies will help increasing the productivity comes from water use and this enables consumers to pay, at least partially, tariffs and accelerate the economic cycle.

10.6 Ethics and Water

The notion of water ethics has raised since the past two decades where it was catalyzed by environmental concerns and the increased demand for water besides abrupt water deficit in many regions worldwide. Hence, the significance of ethics in water use becomes a proposed solution in many cases. Nevertheless, taking actions to improve water ethics is still a real challenge.

Ethics is a socially accepted moral standard to define what can be done and what cannot do done, and/or a standard of what harm or pain, such as damage, loss, poverty, thirst, etc. (UNESCO 2011). Thus, self-responsibility is the regulator for the wise-use of water, better management and conservation approaches of water resources.

The World Commission on the Ethics of Science and Technology (COMEST) stated that, rather than analyzing the ethical issues of water management, it should try to promote best ethical practices. They identified some fundamental principles, including mainly human dignity, participation, solidarity, human equality, stewardship, transparency, inclusiveness and empowerment (COMEST and IHP 2004).

However, the specific added value of ethical considerations lies in providing the orientation for ongoing debates on water challenges by not only applying substantial principles, but by offering suitable procedures as well (Grunwald 2016).

10.6.1 Aspects of Bad Ethics in Water

There are many examples that can be illustrated to show the non-ethical behaviour towards water in Lebanon. For example (but not limited) if water pollution in Lebanon is viewed on the individual level, thus approximately 20% of supplied water for household uses is lost due to unwise use of water, whereas this percentage is higher in case of irrigation purposes and may reach 45%. Whereas, if moral standards are considered by the people who caused such water loss; therefore, the percentage of water loss can be reduced (for example) to 5% and 10% in the household and irrigation uses; respectively.

Perhaps there are several examples mentioned in this book about the bad (non-ethical) actions of people towards water resources in Lebanon including surface and groundwater, but these actions can be categorized into different levels as follows:

1. Actions on individual level:

Bad behaviour of individuals on water differs from one region to another in Lebanon. Perhaps the reason behind this, is the aspects of water shortage in each region as well as the available of water resources there. However, the most common unfavourable actions are:

– Careless use of domestic water where sometimes the amount of water lost exceeds the amount of used water,
– Excessive water use with high flow rates in several utilities (e.g. car washing, construction materials production, etc.),
– Neglect water flow even after filling storage reservoirs and ponds. This is known in both rural and urban areas,
– Chaotic irrigation methods (e.g. furrow and flood irrigation are the most applied methods),
– Chaotic water abstraction, with unorganized approaches, from boreholes, rivers, springs and lakes,
– Dumping different types of liquid and solid wastes in river courses,
– Excessive use of fertilizers in cultivated areas,
– Withdraw water illegally from the public water pipelines.

2. Actions on institutional level:

   – Inequity in water supply from public water sources to consumers for financial or favouritism reasons,
   – Ignoring making maintenance and functioning the water supply systems and the related operations, which is usually conditioned by receiving money,
   – Counterfeit and manipulate water bills of consumers for private benefits.

3. Actions on professional level:

It is also the responsibility of professionals to behave ethically with issues related to water. In this respect, there are many people in Lebanon who work on managing and studying water resources at different levels and disciplines, even though they are not specialized on the subject matter. This has been well pronounced in many institutions IN Lebanon (e.g. academic, public institutions, NGOs, etc.) where unspecialized people (e.g. employee in water sector, lecturer, engineers, etc.) carry out works, responsibilities and studies on water resources, and then they give measures and estimations but with erroneous results and lack to the reliability.

This actually resulted fake information and estimations on water measures and assessment in Lebanon, and therefore, much chaos has been created. It is a sort of corruption where political favoritism occupies a role by putting inappropriate people in the wrong position.

According to the socioeconomic survey applied for this study, it has been found that people who work on water resources including studies, assessment, management and even project execution in Lebanon, are less than 12% specialized in water fields (e.g. hydrology, hydrogeology, water engineering, water biology, etc.). Whereas the largest number of people are (as an example), as follows: 34%, 31%, 17% and 6% for civil engineering, basic sciences, agronomy and miscellaneous; respectively.

10.6.2 Changing Ethics in Water

Perhaps, changing ethical behaviour is much more difficult than taking measures in implementation water resources management. This is because changing ethics in water, according to Priscoli et al. (2004) requires: (a) the acceptance of a set of moral principles, (b) the personal perception of the factual situation and (c) a derived moral judgment on the particular case under consideration.

Therefore, changing ethical behaviour will be self-action depends mainly on the intention of individual him/herself.

Based on the bad actions on the levels illustrated previously, the following can summarize the measures to be taken to regulate behaviours towards water:

1. Encouraging people to try resolve water problems. According to Callahan (1988), it would be profitable to encourage groups to take ethical problems relating to water within their ambit,

2. New technologies (e.g. desalinization, genetic foods, dry com-posting, etc.) with their utility in saving and supply sufficient amounts of water will change how we look at water,
3. Awareness and capacity building would play essential role in establishing knowledge on water availability and scarcity, and this in turn will regulate the ethics towards water,
4. Empower women involvement in water management, since de facto, women are the key water user and self-manger in many societies. Nonetheless, women are rarely involved in strategic decision-making processes regarding water resources management,
5. There must be actions taken towards ethically regulating the adoption of people who work in water studies and employments,
6. Adopting legislations and policies accompanied with punishment approaches taken for people who unethically act towards water.

10.7 Lebanon in SDG

As a global initiative, the Inter-Agency Expert Group on SDGs developed Global indicators for the sustainable development goals (SDGs) which were subsequently adopted by the United Nations Statistical Commission in March 2016. This includes seventeen goals with 244 indicators where Goal 6 aims at: *ensuring availability and sustainable management of water and sanitation for all.*

The engagement of Lebanon in this global activity is useful for better orientation of the existing policies and plans as well as it helps finding new methodologies and approaches for the monitoring systems, notably that the issue of water and sanitation data and monitoring are still problematic in Lebanon, and this needs a guidance by international agencies who have been involved in such programs and can disseminate lessons learned to reinforce the country capability for securing it water and sanitation systems through successful integrated management approaches.

Upon the request of a group of custodians including (UN Environment, WHO, UNICEF, UN-Habitat, UNSD, UNESCO, UNECE FAO and OECD, Lebanon, through the CNRS-L (as the IHP Focal Point) shared in reporting data for six indicators related to SDG-6. These are: SDGs 6.2.3, 6.4.1, 6.4.2, 6.5.1, 6.5.2 and 6.1.1.

The significance of SDG 6 for Lebanon differs between the illustrated indicators, and this is based on the existence of the issue of each indicator and the degree of impact on water and sanitation sectors, as well as the capability to elaborate the proper solutions is also significant. This can be summarized as in Table 10.4.

It is obvious form Table 10.4 that the six elaborated indicators on water and sanitation pointed out to unfavorable situation for Lebanon. Thus, a need to take action becomes a priority in this respect, and this requires many integrated implements to be taken together to secure water and sanitation systems (Shaban et al. 2018).

Integrated monitoring can be one of these actions which must be taken as a first step where inventory on data and information must be prepared and supplemented

Table 10.4 SDG6 indicators, with elaborated surveys, and their applicability to Lebanon (Shaban et al. 2018)

Indicator	Description	Related problem in Lebanon	Degree of negative impact	Capability for solution
6.2.3	Water bodies of water with good ambient water quality	Well pronounced	High	Needs systematic monitoring control
6.4.1	Change in water use efficiency over time	Increased water stress	High	Institutional coordination and awareness are needed
6.4.2	Freshwater withdrawal as a proportion of available freshwater resources	High level of freshwater withdraw	High	Legislations and environmental controls must be elaborated
6.5.1	Integrated water resources management implementation	Poorly managed	Moderate	There is a need for new policies
6.5.2	Transboundary basin with an operational arrangement for water cooperation.	Few/weak treaties	Moderate	Updated treaties should be followed.
6.6.1	Change in the extent of water-related ecosystems over time	Well pronounced	Moderate to high	Institutional coordination and capacity building are needed

by continuous monitoring approaches using different tools and methodologies. This can be also supported by adopting successful approaches applied in many regions and proved their utility in the enhancement and consolidation of water and sanitation sectors.

Chapter Highlights
- The author declared that "Water in Lebanon is in Jeopardy."
- The spatial and temporal dimensions of water crisis in Lebanon has been well diagnosed and creditable estimations were put.
- The future of water resources in Lebanon has been viewed and predication were illustrated based on the discussed spatial and temporal dimensions.
- It is anticipated that Lebanese rivers will lose their water totally after 125 years, while springs need 115 years.
- Water discharge from wells will be decreased by 1% per years, and it will be reduced at 10% by 2030.
- There is about 20 m³/capita will be decreased from water quota of the inhabitants in Lebanon by 2030.
- There are several solutions proposed to reduce the impact of the existing challenges on water resources in Lebanon. Hence, different aspects of water harvesting are considered as the principal and most optimal solution.

- Artificial Groundwater Recharge (AGR) can be one of the most successful technology to be applied, notably in the coastal zone.
- Tapping groundwater discharges into the sea was discussed in-depth.
- Proposed solutions to reduce the degree of water contamination have been put.
- There are effective economic policies proposed as a contributing solution to enhance the status of water sector in Lebanon.
- Ethics and water was discussed in details. Thus, many aspects of moral behaviors were recommended in order to reach the most optimal wise use of water.
- Mainstreaming of SDG-6 in the water strategies and policies in Lebanon was tackled, and thus the applicability of the SDG-6 indicators was illustrated.

References

Al Saud M (2015) Flood control management for the city and surroundings of Jeddah, Saudi Arabia. Springer, New York/London, 169p

Ayoub G, Khouri R, Ghannam J, Acra A, Hamdar B (2002) Exploitation of submarine springs in Lebanon: assessment of potential. J Water Supply Res Technol AQUA 51(1):47–64

Bakken T, Ruden F, Mangset L (2012) Submarine groundwater: a new concept for the supply of drinking water. Water Resour Manag 26:1015–1026. https://doi.org/10.1007/s11269-011-9806-1

Berbel J, Gutiérrez-Martín C, Martin-Ortega J (2017) Water economics and policy. Water MDPI Publishing 9:801. https://doi.org/10.3390/w9100801

Callahan J (1988) Ethical issues in professional life (Callahan JC (ed)). Oxford University Press, New York, pp 3–25, 470 pp

COMEST (World Commission on the Ethics of Science and Technology) IHP (International Hydrology Programme) (2004) Best ethical practice in water use. UNESCO, Paris

ESCWA (United Nations Economic and Social Commission for Western Asia) (2010) Water sector in Lebanon an operational framework for undertaking legislative and institutional reforms, 46pp

FNC (First National Communication) (2009) Climate change. Ministry of Environment & GEF & UNDP, Beirut, 164pp

Grunwald A (2016) Water ethics – orientation for water conflicts as part of inter- and transdisciplinary deliberation. In: Society – water – technology. Springer, Berlin, 11–32pp

Hamzé M, Shaban A (2018) Conclusion and discussion. In: The Litani River, Lebanon: an assessment and current challenges. Springer, Cham, 179p

Hassanli A, Beecham S (2009) Criteria for optimizing check dam location and maintenance requirements. In: Check dams, morphological adjustments and erosion control in torrential streams. Nova Science Publishers, New York, 1–22pp

Kashoua L (2015) Rooftop water harvesting: an adaptation instrument for water shortage in Lebanon. MSc dissertation, Lebanese University, 49pp

Kato T, Endo A (2017) Contrasting two dimensions of disaster-induced water-shortage experiences: water availability and access. Water Water 9:982. https://doi.org/10.3390/w9120982

Kraemer A, Castro Z, Seroa de Motta R, Russell C (2003) Economic instruments for water management: experiences from Europe and implications for Latin America and the Caribbean. Integration and Regional Programs Department Sustainable IADB, New York, 83pp

Masciopinto C (2013) Management of aquifer recharge in Lebanon by removing seawater intrusion from coastal aquifers. J Environ Manag 130:306–312

Mhawej M, Faour G, Fayad A, Shaban A (2014) Towards an enhanced method to map snow cover areas and derive snow-water equivalent in Lebanon. J Hydrol 513:274–282

MoEW and UNDP (2014) Assessment of groundwater resources of Lebanon, 88pp

Nyssen J, Veyret-Picot M, Poesen J, Moeyersons J, Mitiku H, Deckers J, Govers G (2004) The effectiveness of loose rock check dams for gully control in Tigray, northern Ethiopia. Bed Sediment Use Manag 20:55–64

Priscoli J, Dooge J, Llamas R (2004) Water and ethics. United Nations Educational, Scientific and Cultural Organization, Paris, 33pp

Shaban A (2003) Etude de l'hydrogéologie au Liban Occidental: Utilisation de la télédétection. PhD dissertation, Bordeaux 1 Université, 202p

Shaban A (2016) New economic policies: instruments for water management in Lebanon. Hydrol Curr Res 7(1):1–7

Shaban A, Darwich T (2008) Assessment of hill ponds sites in Ar-Rssal area. Unpublished Technical Report. Development Studies Association, 19pp.

Shaban A, Dawrich T (2011) The role of sinkholes in groundwater recharge in mountain crests of Lebanon. Environ Hydrol J 19(9):2011

Shaban A, Khawlie M, Abdallah C, Faour G (2005) Geologic controls of submarine groundwater discharge: application of remote sensing to North Lebanon. Environ Geol 47(4):512–522

Shaban A, Robinson C, El-Baz F (2009) Using MODIS images and TRMM data to correlate rainfall peaks and water discharges from the Lebanese coastal rivers. J Water Resour Prot 4:227–236

Shaban A, Darwich T, Drapeau L, Gascoin S (2014) Climatic induced snowpack surfaces on Lebanon's mountains. Open Hydrol J 2014(8):8–16

Shaban A, Faour G Mhawej M (2018) Developing institutional capacity for integrated approach in SDG 6 monitoring in Lebanon. Technical report submitted to UN-Water, 41p

SNC (Second National Communication) to the UNFCCC (2011) Climate change vulnerability and adaptation. Ministry of Environment & GEF & UNDP, Beirut, 288 pp

TNC (Third National Communication) to the UNFCCC (2015) Updated national greenhouse gas (GHG) inventory, mitigation, vulnerability and adaptation analysis and policy recommendations Ministry of Environment & UNDP, Beirut (Under preparation)

UNEP-IETC (2001) Sourcebook of alternative technology for freshwater augmentation in West Asia. Technical publication series (8 f). UNEP-IETC, Osaka, 182 pp

UNESCO (2011) Water ethics and water resource management. Ethics and climate change in Asia and the Pacific: Working Group 14. Report. Regional Unit for Social and Human Sciences in Asia and the Pacific, Bangkok, 84pp

UN-HABITAT (2008) Country programme document 2008–2009, Lebanon. Available at: http://www.unHABITAT.org/pmss/listItemDetails.aspx?publicationID=2706

UNHCR (High Commissioner for Refugees) (2019) Lebanon water sector end of year dashboard 2018. Available on: https://reliefweb.int/report/lebanon/lebanon-water-sector-end-year-dashboard-2018

Varis O, Keskinen M, Kummu M (2017) Four dimensions of water security with a case of the indirect role of water in global food security. Water Scarcity 1:36–45

WB (World Bank) (2012) Lebanon country water sector assistance strategy 2012–2016. Sustainable Development Department Middle East and North Africa Region. Report no. 68313-LB, 36p

Xuan Y, Holly M (2019) Offshore pumping impacts onshore groundwater resources and land subsidence. Geophys Res Lett 46(5):2553. https://doi.org/10.1029/2019GL081910. Terrain Surface Cracking

Index

A

Abstraction, 134, 148, 155, 180, 199, 208, 212
Accumulation/melting, 77, 78
Accumulations, 27, 75, 78, 81, 82, 86, 91,
    110, 113, 170
Adaptation, 94, 126, 162, 167, 178, 198, 210
Aerial photographs, 40, 75
Afqa Springs, 39, 60, 62
Africa, 2, 109
Aggregates, 100, 103, 200
Agricultures, 6, 9, 15, 17, 18, 30, 110, 170,
    172, 210
Air masses, 2, 10
Al-Assi River, 2, 36, 40, 43, 45, 47, 49, 50
Alluvial deposits, 26, 28, 113, 149, 175,
    205, 207
Alternative source, 196
Altitude-related, 17, 19
Ammiq wetlands, 111, 118
Ancient, 2, 3, 92
Annual discharge, 36, 50, 51, 61–63, 177
Annual rainfall, 4, 16, 18, 49, 169
Anthropogenic challenges, 181
Anticlines, 25, 27–30, 101, 148
Anti-Lebanon, 2, 17, 20, 25, 27–29, 74, 97,
    134, 168
Aquiclude, 127, 129, 131, 150
Aquiferous, 4, 127–131, 134, 148–152, 204
Aquiferous rocks, 109, 126–129, 132–136,
    148, 154–157, 175, 186, 187,
    189, 200–203
Aquifers, 6, 40, 48, 56, 60, 64, 75, 126–136,
    148–154, 156, 168, 175–177, 179,
    187–189, 191, 199, 200, 203, 204, 208

Aquitard, 127, 129, 130, 152, 155
Arable lands, 186, 188, 190–192
Arab Region, 2, 199
Arc-GIS, 138, 144
Archeological sites, 115
Argillaceous, 25–27, 56, 83, 99,
    101, 129, 139, 151,
    193, 195, 199
Aridity, 18–20
Aridity index, 18, 19
Artesian springs, 60, 68, 113
Artificial groundwater recharge (AGR)
    technology, 198, 218
Artificial lakes, 90, 91
Artificial recharge, 126, 198
Asia, 2
Aster, 41, 137, 138, 143, 177, 204
Atmosphere, 13–31, 84, 171
Awareness, 216, 217

B

Bare lands, 30, 139
Beddings, 29, 54, 66, 113, 116, 135, 137, 199,
    202, 205–207
Beirut, 3, 4, 8, 9, 39, 43, 45, 47, 48, 74, 100,
    102, 155, 166, 174, 176
Bekaa Plain, 2, 6, 17, 18, 20, 21, 25, 28,
    36, 39, 40, 56, 94, 97, 134,
    149–153, 168, 180
Bifurcation ration, 138, 194
Biological features, 112, 114
Bisri Dam, 100–102, 104
Borders, 2, 20, 49, 75, 177